NATURALLY NANTUCKET

THE SEA: VOLUME 1

SARAH D. OKTAY

Naturally Nantucket The Sea: Volume 1 is compiled of various online and in print essays published by Yesterday's Island (https://yesterdaysisland.com/) written by the author between 2008-2016. Rights transferred to the author.

ISBN: 978-1-7341378-2-8

Published in the U.S. by Viola Road Press

Dedication

For my friend, Peter H. McCartney, a man who loves both the sea and science, in appreciation for his support of the Organization of Biological Field Stations and field stations and marine labs everywhere. Peter has always been supportive of field station directors like me, who struggle daily to provide a safe and well-equipped remote science home. OBFS field stations and marine labs are scattered all over the globe, serving as a scientific remote outpost, like a microscope-equipped Airbnb to allow students and scientists to live and work in locations from the poles to tropical jungles. Thank you, Peter!

This book of essays is also dedicated to my husband, the poet Leonard Germinara, who inspired me to write down what I saw and figure out what I needed to know.

Author's Note and Acknowledgments

This book came about for a variety of reasons while taking a long and winding path that I never expected to be traveling. My journey started when I became the Managing Director of the University of Massachusetts Boston's Nantucket Field Station (UMB-NFS) in 2003. The Nantucket Field Station consisted of a laboratory, dorm space for 14, a garage, and an upstairs apartment where my husband and I lived on 107 acres of harbor front property on Nantucket Island, 26 miles off the southern coast of Cape Cod in Massachusetts. The Field Station also included housing for 26 students in the form of old U.S. Navy condos in the mid-town part of the island. For twelve years, I was fortunate to run the UMB-NFS which operated year-round for scientists and students of all ages. By training, I am a chemical oceanographer and when possible, I conducted water quality, oceanographic, and coastal processes research with citizen scientists and high school and college interns on island. I rubbed elbows with celebrities and politicians and rich and poor and in the end, was richer for the experience.

Back in 2007, my friends Suzanne Daub and the sadly departed Jerry Daub asked me to start writing a column for their seasonal Nantucket Island publication, *Yesterday's Island*. For the next eight years I would write approximately

22 columns each year on a subject of my choice, which typically was on whatever science or nature related cool thing was occurring at the field station that week. I essentially wanted to research a variety of topics so that I would understand them better when people asked me about them.

Once I realized I was looking at about an 800-page book when all my essays were combined, I decided to split this work into four volumes, entitled the Sea, Volumes 1 and 2 and the Land, Volumes 1 and 2. I resigned myself to recognizing my verbosity which has been self-evident to anyone who's ever known me going back to my elementary school teachers. I was not one of those kids who agonized about reaching a 500- or 1000-word count or three-page goal. Typically, I would go over, sometimes way over the page count. Once I started the task of piling topics into "Land" or Sea," I found many salty connections which I lumped into these first two volumes as most people interested in the ocean are also interested in salt marshes and coastal processes. In this volume, we will investigate basic island coastal processes and weather events that dominate island dynamics. Then we will meander through articles on a variety of sea life, concentrating on invertebrates like clams and crabs. In Volume II of the Sea, I will concentrate on articles about sharks, seals, and whales and describe some of the crazy and little seen creatures that wash up on Nantucket beaches. Next, will be ocean-adjacent species of birds and plants that tend to be found in coastal regions. Finally, we will end up with odds and ends as well as some advice on how we can better care for our planet. You'll find in "The Land" Volumes 1 and 2, articles on cattails and phragmites (common reed) that are often associated with marshes but are essentially freshwater dependent plants. Island biogeography articles are also featured in "The Land." Of course, with small islands it is very hard to separate the influence of the ocean.

I want to apologize right up front. I have done my best to change the tense to past tense when I could, but these were written as adventures occurred and in response to marine mammal strandings, weather events, and seasonal natural events. These are meant to be conversational and informative, and I am sure I will fail more than once trying to translate the science in a fun way.

I could not have written this without my patient husband, Leonard Germinara, who endured eight years of me working through the night every Sunday to make my Monday morning deadlines. Len ran a stellar junior ranger program at the Field Station with youngsters serving as docents and nature guides. His ability to teach observational skills and instill a deep sense of wonder and excitement are inspirational. He also incorporated the arts via drawing, poetry, photography, cartooning, and writing with his students. You can find their work at http://jrrangerfieldnotes.blogspot.com/.

I have been blessed to have a loving family, my brothers Alan and Jaime are near and dear to me and extremely supportive of their big sister. My sister-in-law Stephanie never fails to amaze me. To my friends on Nantucket, no matter where I travel throughout the world, I'll never find a more lovable group of intelligent misfits. I'd also like to dedicate this to my many friends in the Organization of Biological Field Stations. I have never met a more interesting and altruistic group of "blue-collar" scientists.

Table of Contents

Section 1: Coastal Processes and Habitats, Storms, and Weather

Living with the Sea
(Oct 1, 2009)

The fall equinox occurred just a few days ago (September 22) and winter will soon descend upon the island. One of the pleasures of living on Nantucket is the opportunity to become more attuned with the seasons, the night sky, the rise of the sun and moon, and the tides. A review of some of the celestial and meteorological phenomena we've experienced lately seems fitting as the days wind down and darkness creeps earlier and earlier into our daytime schedules.

Of course, I must start an article about tides with a quote I have always liked, "time and tide wait for no man." This phrase is a *figura etymologica*[1] in which two similar words are used together for emphasis, in this case, both time and tide derive from the Proto-Germanic words tīmô and tīdiz which are essentially the same word. The word tide originally referred to the passing of time, not what we see at the seashore. This expression is very old, the earliest known record is from St. Marher, in 1225 in Latin: "And te tide and te time þat tu iboren were, schal beon iblescet."[2] While we are learning our Latin, I bet you might not know the actual story of King Canute also known as Cnut the Great, the King of Denmark, England and Norway (born 995ish, died 1035). King Canute's story is often mistold as the story of an arrogant king who tells his couriers to place his throne in the surf and commands the rising tide to not wet his feet. In the original story as recorded in the 12th century by Henry of Huntingdon in *Historia Anglorum* (The History of the English People) King Canute is demonstrating to his subjects, that even as a King, he could not stop the tide and that only God could command the elements.[3]

Before I accidentally begin a thesis to obtain either a history or literature degree, let's get back to the original intent of this article. What are tides and how do our seasons change in relationship to the height of the sea? Whether one is tying up a boat, surf casting, driving out to Great Point, or planning a beach stroll, it pays to be aware of the tides. Tides are the product of the periodic motion of the waters of the sea due to changes in the attractive forces of the moon and sun upon the rotating earth. Water quite literally sloshes back and forth as the earth turns on its axis and a gravitational tug is exerted by the moon and the sun. On top of that a gravitational pull is exerted by the earth itself, slightly complicating things. And of course, the earth is spinning on its axis. During the new and full moons, the distance between the high and low tides increases, which means that high tides get higher and low tides get lower. These are called spring tides and they produce on average about twenty percent more amplitude in the high and low tides. On Nantucket, the typical maximum difference between high and low tides is about four feet.

In most parts of the world, the difference between high and low tide amounts to only a few feet. Nevertheless, this change in water level can be particularly important in some areas. This is painfully obvious when attempting to launch a boat at low tide. Because of their effect on navigation, tides have been studied for many years, and today physical oceanographers know quite a bit about their causes and characteristics. As we will learn, sometimes anomalies occur that force us to rethink some of our assumptions and may encourage us to do a better job of recording our local tides.

This summer (2009) you might have noticed that Nantucket experienced higher than normal high tides in June and July. Many of the tidal cycles we saw were of the magnitude one would expect during winter, when the earth is tilted closer to the moon. Here at the UMASS Boston Nantucket Field station, we can observe the tides daily, and I was especially surprised to see the abnormally high tide levels. My first clue was the fact that I had to practically tread water during our late-night horseshoe crab inventory walks in the surf, when typically, the water is only waist deep. Fortunately, we were not imagining things; the

National Ocean and Atmospheric Administration (NOAA) and many Atlantic coast scientists, harbormasters, and fishermen recorded the same phenomenon.

"Starting in early June 2009, observed tides have been increasingly elevated above predicted tidal elevations along the entire U.S. East Coast from Maine to the east coast of Florida. During the period from June 19 through June 24 for instance, these water levels were running between 0.6 to 2.0 feet above normal depending upon location. As of July 1, these anomalies continue, but running lower at 0.3 to 1.0 ft. above normal. It is not unusual for smaller regions and estuaries along the U.S. East Coast to experience this type of anomalous event at this time of year; however, the fact that the geographic extent of this event includes the entire East Coast event is anomalous."[4]

NOAA released a technical report that addresses some of the theories for why this anomaly might have occurred. "'The ocean is dynamic and it's not uncommon to have anomalies,' said Mike Szabados, director of NOAA's Center for Operational Oceanographic Products and Services. 'What made this event unique was its breadth, intensity and duration.'"[5]

The NOAA report notes that some of the highest sea levels occurred closer to where the anomaly was thought to have formed in the Mid-Atlantic. Cities like Baltimore, Maryland experienced extreme high tides as much as two feet higher than normal. After observing water levels six inches to two feet higher than originally predicted, NOAA scientists began analyzing data from select tide stations and buoys from Maine to Florida and found that a weakening of the Florida Current Transport—an oceanic current that feeds into the Gulf Stream—in addition to steady and persistent Northeast winds, contributed to this anomaly. Whether this is a decadal or cyclic phenomenon or is somehow related to the El Nino/La Nina cycles remains to be seen. Scientists are certainly going to be paying more attention to the east coast tides.

One of the items mentioned in the report is that the impacts of the event were amplified by the occurrence of a perigean-spring tide, which occurs close to the vernal equinox when the moon is closest to the Earth and its gravitational pull heightens the elevation of the water. The combined effects of this tide with the sea level anomaly produced minor flooding on the coast.

In the language of science, an equinox is either of two points on the celestial sphere where the ecliptic and the celestial equator intersect. For the rest of us, it's one of two times a year when the Sun crosses the equator, and the day and night are of approximately equal length. For many people the autumnal equinox is a sign of fall and a harbinger of cool nights and approaching holidays.

"Equinox" means literally, "equal night." As the angle of the earth's inclination toward the sun changes throughout the year, lengthening or shortening the days according to season and hemisphere, there are two times annually when day and night are of roughly equal duration: the spring and autumnal equinoxes. These celestial "tipping points" have been recognized for thousands of years and have given rise to a considerable body of seasonal folklore.

At the autumnal equinox (Sept 22, 2009; 5:18 P.M. EDT), the Sun appears to cross the celestial equator, from north to south; this marks the beginning of autumn in the Northern Hemisphere. The vernal equinox, also known as "the first point of Aries," is the point at which the Sun appears to cross the celestial equator from south to north. This occurs about March 21, marking the beginning of spring in the Northern Hemisphere. The equinoxes are not fixed points on the celestial sphere but move westward along the ecliptic, passing through all the constellations of the zodiac in 25,800 years. This motion is called the precession of the equinoxes.

Hopefully, now that fall is here, the island can begin to dry out. According to the National Climate Report for August 2009 by NOAA[6], Massachusetts received 152% of normal rainfall since January, resulting in the third wettest summer in Massachusetts through 2009! Although rain can be a damper, it is rarely a significant factor in local tides. Large rainfall events can set up more stratified conditions in our ponds and harbor and is often the culprit in causing huge algal blooms as the excess water washes more material into our ponds. After a heavy rainfall, the harbors often must be closed to shellfishing because of a relationship found between this runoff and increased bacterial production. Occasionally, an excess of fresh water can form a lens on saltwater if mixing by winds does not occur. That lens can allow algae to grow faster simply through the amplification of sunlight in the water column which is exacerbated by the extra nutrients washed into the harbor.

Our next stop will be winter, when we brace for storms, higher tides, and beach eroding winds. For now, I am going to enjoy my favorite time on the island and try to remember to pay attention to the constantly changing forces of the land and sea.

Winter Weather: An Island Blanketed by the Sea
(November 14, 2013)

I am writing this article on an unseasonably warm Sunday in November. According to Weather Underground which is usually my go-to source for weather data, the maximum temperature was 57 °F on Nantucket on the

10th of November of this year, 2013.[7] Just three days ago, it was over 70 °F on Nantucket and hit 68 °F in Boston, missing the record high of 77 °F recorded in 1938[8] and much higher than the 54 °F average for that date in Boston. Today (the 10th) the ocean temperature surrounding the island is 50 °F. Our comfortably warm fall days are a result of the ocean surrounding the island that provides a buffering envelope of water that keeps us warmer in the winter and cooler in the summer. The layer of insulating water provides about a 10 °F difference in temperature for Nantucket versus mainland Massachusetts. Some Wampanoag took advantage of the cool breezes that swept the island each summer, abandoning their mainland encampments in the summer to come to the island to cool off. This summer cooling effect is due to the ocean and our distance from a large land mass. The Gulf Stream's warm waters reach close enough to provide warm breezes and strange creatures, while the surrounding water prevents Nantucket from becoming too hot when the mainland bakes each summer.

The Köppen climate classification[9] system is a commonly used system for assigning climate characteristic based on the types of plants that can grow in certain regions as a result of temperature, humidity, rainfall and other factors. It was created by the German botanist-climatologist Wladimir Köppen. His aim was to devise formulas that would define climatic boundaries in such a way as to correspond to those of the vegetation zones (biomes) that were being mapped for the first time during his lifetime. Köppen published his first scheme in 1900 and a revised version in 1918. The Köppen climate classification system divides climates into five main climate groups: A (tropical), B (dry), C (temperate), D (continental), and E (polar). The second letter indicates the seasonal precipitation type, while the third letter indicates the level of heat. According to the Köppen climate classification system, Nantucket features a climate that borders between a humid continental climate (Dfb) and an oceanic climate (Cfb), the latter a climate type rarely found on the east coast of North America. Nantucket's climate is heavily influenced by the Atlantic Ocean, which helps moderate temperatures here throughout the year. As a result, the island's winter climate is warmer than that on the mainland of New England, and summers are cooler than on the mainland. Average temperatures during our coldest month (January) are just below 32 °F (0 °C), while average temperatures during our warmest months (July and August) hover around 69 °F (21 °C). Our tourism success and appeal are in no small part linked to the moderate weather and cool summers.

Nantucket receives on average 38 inches (970 mm) of precipitation annually, spread relatively evenly throughout the year. Like other cities with an oceanic climate, Nantucket experiences many cloudy and overcast days, particularly outside the summer months. Our yearly snowfall totals are much less than the mainland which was very evident in the winter of 2012/2013 with four feet plus on the mainland and less than a foot of snow here.

One of the reasons for this differential heating between water and land is that water is a transparent medium and land is opaque. Water allows light to penetrate to depth, leaving the surface layers cooler than they would be if the surface was opaque. A cooler water surface results in cooler air temperatures above. When solar radiation strikes land, the energy is absorbed in a thin layer that heats relatively rapidly. Likewise, it readily gives up its heat to the atmosphere. In the center of the continents the largest temperature swings occur.

The ocean is our planet's largest heat sink. By absorbing, storing and then slowly releasing large quantities of heat, the ocean buffers the climate of the nearby land and, over time, the entire planet. The very slow heat release of the ocean allows it to transport this heat all over the world. The heat carried by the Gulf Stream water travels over to Europe and keeps England and the Mediterranean parts of Europe much warmer than they would typically be at that latitude. This heat balance and the effect that climate change and warming sea temperatures are having means that more heat is being transported, and that the ocean's metabolism effectively is speeding up. One of the biggest concerns relative to our warming atmosphere and oceans is that hurricanes and storms will have more energy to draw from.

Of course, the fact that we are sitting out in the middle of Nantucket Sound with the Atlantic Ocean battering us on the east and the south shores also contributes to our high winds. Winds have plenty of time to build up steam with no mountains or hills or topography to slow them down.

Our much cooler spring temperatures and delayed start for the growing season can also be attributed to the waters surrounding the island. It takes a long time for water temperatures to change. Compared to air or land, water is a slow conductor of heat. That means it needs to gain more energy than a comparable amount of air or land to increase its temperature. Although water is a slow conductor of heat, it tends to store heat quite well. That means that, once heated, a body of water will hold onto that heat for a much longer period of time than either air or land. If you have ever tried to swim early in the season on Nantucket, you will quickly find out that the water surrounding the island typically does not reach sixty degrees until the last few days of June.[10]

Climate change and the direct results of higher ocean sea surface temperatures (SSTs) will accelerate our spring warm up and has already affected our winter temperatures. According to the Environmental Protection Agency[11] and a variety of state and federal sources, the average temperature in the Commonwealth of Massachusetts has increased 2 °F in the last century. By 2100, temperatures could increase by about 4 °F in winter and spring and about 5 °F in summer and fall. Although we have been buffered some here on Nantucket, Massachusetts has warmed almost twice as much as the rest of the contiguous 48 states.[12]

[Note: As you may know, the EPA web sites in 2020 are not the same as they were in 2016, so pay close attention to the end notes here to access working links to critical information.]

Average annual precipitation in the Northeast increased 10 percent from 1895 to 2011, and precipitation from extremely heavy storms has increased 70 percent since 1958.[13] During the next century, average annual precipitation and the frequency of heavy downpours are likely to keep rising. Average precipitation is likely to increase during winter and spring, but not change significantly during summer and fall. Rising temperatures will melt snow earlier in spring and increase evaporation, and thereby dry the soil during summer and fall. So flooding is likely to be worse during winter and spring, and droughts worse during summer and fall. This follows pretty much the same pattern that we are seeing globally.

It has become imperative to ensure that we have set aside money to record these changes and develop predictive models. Research field stations around the world and local, state, and federal agencies have prioritized installing and maintaining weather stations, stream gauges, and offshore data buoys to record these changes and allow scientist to update their models and track trends. It is extremely obvious when you start looking for data on how quickly temperatures and rising and climate effects are occurring, that even predictions and estimates from eight years ago have already been eclipsed. Soapbox alert: Please encourage your local and state and federal representatives to continue to fund data collection and share it with the public.

Fog Happens
(August 25, 2011)

On Nantucket, fog is a normal part of life and trying to predict when to expect it and how long it may last is an essential part of travel to and from the island. In fact, Nantucket acquired one of its most often used nicknames, the "Grey Lady" due to its frequent blanket of fog. If you live here, you eventually learn to love fog or at least appreciate its beauty if you don't have plans to fly off island.

Fog is essentially a cloud on land and occurs when water droplets condense out of moist air because the air temperature is equal to the dew point. It is a simple as that. Ever since I was a child I have been fascinated by clouds and fog. Even now, many decades later I am mesmerized while flying through clouds. My father's PhD thesis was about cloud particulate condensate, so maybe it is genetic! According to the American Meteorological Society (AMS)[14], fog forms when the difference between the temperature and dew point is generally less than 2.5 °C or 4 °F. The dewpoint is the temperature at which the water vapor saturates a parcel of air to the point where it can no longer hold any more moisture and condensation begins. If you think of a crowded nightclub or party, eventually the human density reaches the point where when one more person enters the room, everyone else cannot move about freely and someone ducks out a back or side door. That is essentially like water droplets condensing out of saturated air. Fog can form as the temperature drops a little below its dewpoint producing radiation fog, advection fog, or upslope fog. Or you can add moisture to the air and elevate the dewpoint producing steam fog or frontal fog.

Fog is distinguished from mist only by its density and the amount of terror it causes people. When you think of it, both the Mist and the Fog have been set pieces for horror movies, books, and cemeteries everywhere. No self-respecting spook house can exist without some dry ice (solid carbon dioxide) lying around creating fog by sublimation (the state of going directly from a solid to a gas). But in science land, the difference between mist and fog is in how they affect visibility. Fog reduces visibility to less than one kilometer (5/8 statute mile), mist reduces visibility to no less than one kilometer. So technically, you can see the monster a little sooner if you are in mist versus fog. According to U.S. weather observing practice, fog that hides less than 0.6 of the sky is called ground fog. If fog is so shallow that it is not an obstruction to vision at a height of 6 ft above the surface, it is called simply shallow fog. Because I am 5 feet 2 inches on a good day, they both would seem to be the same to me.

Haze is traditionally an atmospheric phenomenon in which dust, smoke, and other dry particulates obscure the clarity of the sky. It is in effect a "dry fog" created when fine particulates get suspended in the air. Fog is easily distinguished from haze by its higher relative humidity (near 100%) and gray color. Mist may be considered an intermediate between fog and haze; its particles are smaller (a few micron, or 10^{-6} meters μm maximum) in size, it has lower relative humidity than fog, and does not obstruct visibility to the same extent. As we all have learned since the beginning of the industrial age, in industrial areas and in big cities, fog is often mixed with smoke, and this combination has been known as smog. Haze is also sometimes referred to as smog although it is typically created by photochemical reactions in the air.

Let's dive a little deeper into some of the types of fog you'll see on island. As you navigate the island, you'll see almost every type of fog that exists. Radiation fog is a type of fog that forms at night under clear skies with calm winds when heat absorbed by the earth's surface during the day is radiated into space. As the earth's surface continues to cool, provided a deep enough layer of moist air is present near the ground, the humidity will reach 100% and fog will form. Radiation fog varies in depth from 3 feet to about 1,000 feet and is always found at ground level and usually remains stationary. This type of fog can reduce visibility to near zero at times and make driving very hazardous. This is usually the type of fog we are most familiar with, you know, the kind hanging out in graveyards.

Valley fog is a type of radiation fog that is very common in the humid mountainous regions such as the hills of eastern Kentucky. When air along ridgetops and the upper slopes of mountains begins to cool after sunset, the air becomes dense and heavy and begins to drain down into the valley floors below. As the air in the valley floor continues to cool due to radiational cooling, the air becomes saturated and fog forms. Valley fog can be very dense at times and make driving very hazardous due to reduced visibility. This type of fog tends to dissipate very quickly once the sun comes up and starts to evaporate ("burn off") the fog layer. I've seen a version of valley fog occur as cooling air gets heavy and rolls off the cliff by the field station onto the beach below.

A type of fog we frequently encounter on island is advective fog which occurs when warm moist air moves over a cold patch. Sea fogs are always advection fogs, because the oceans don't radiate heat in the same way as land and so never cool sufficiently to produce radiation fog. Fog forms at sea when warm air associated with a warm current drifts over a cold current and condensation takes place. Sometimes such fogs are drawn inland by low pressure, as often occurs on the Pacific coast of North America. Advection fog may also form when moist maritime, or ocean, air drifts over a cold inland area. This usually happens at night when the temperature of the land drops due to radiational cooling.

Upslope fog forms when light winds push moist air up a hillside or mountainside to a level where the air becomes saturated and condensation occurs. This type of fog usually forms a good distance from the peak of the hill or mountain and covers a large area. Upslope fog occurs in all mountain ranges in North America. This usually occurs during the winter months, when cold air behind a cold front encounters a "wall", let's say the impressive slopes of the Rocky Mountains. As the cold, moist air rises up the slopes of the mountains, condensation occurs and extensive areas of fog form on the lower slopes of the mountains.

Ice fog: This type of fog forms when the air temperature is well below freezing and is composed entirely of tiny ice crystals that are suspended in the air. Ice fog will only be witnessed in cold arctic-like or polar air. It has happened a few times on Nantucket. Generally, the temperature must be 14 °F or colder for ice fog to occur. Freezing fog occurs when the water droplets that the fog is composed of are "supercooled". Supercooled water droplets remain in the liquid state until they encounter a surface upon which they can freeze. As a result, any object the freezing fog touches will become coated with ice.

During the fall on Nantucket, we may experience "evaporation" or "mixing fog". This type of fog forms when enough water vapor is added to the air by evaporation and the moist air mixes with cooler, relatively drier air. The two common types are steam fog and frontal fog. Steam fog forms when very cold air moves over warm water. When the cool air mixes with the warm moist air over the water, the moist air cools until its humidity reaches 100% and fog forms. This type of fog takes on the appearance of wisps of smoke rising off the surface of the water which gives it a common name of "sea smoke". This type of fog is more common in the Arctic and Antarctic[15], but Nantucket gets cold enough for this phenomenon to happen occasionally. The other type of evaporation fog is known as frontal fog. This type of fog forms when warm raindrops evaporate into a cooler drier layer of air near the ground. Once enough rain has evaporated into the layer of cool surface, the humidity of this air reaches 100% and fog forms.

Some of the foggiest parts of the world are very nearby on the Grand Banks off the island of Newfoundland, which is the meeting place of the cold Labrador Current from the north and the much warmer Gulf Stream from the south. Some of the foggiest land areas in the world include the little town of Argentia, Newfoundland and Point Reyes, California, each with over 200 foggy days per year[16]. Even in generally warmer southern Europe, thick fog and localized fog is often found in lowlands and valleys, such as the lower part of the Po Valley and the Arno and Tiber valleys in Italy or Ebro Valley in northeastern Iberia, as well as on the Swiss plateau, especially in the Seeland area, in late autumn and winter. Other notably foggy areas include coastal Chile (in the south), coastal Namibia, and the Severnaya Zemlya islands.[17]

You might recall a July 29th 2010 article in Yesterday's island[18] when we discussed frost pockets in a desperate attempt to cool off by thinking cool thoughts. Frost pockets lead to "foggy bottoms" which exist around Nantucket and are simply cooler depressions in the ground where fog can collect as the cold moist air is denser than the surrounding warmer air. Perhaps you have noticed on the subway (the Metro) in Washington DC the "Foggy Bottom" stop which is by the George Washington University. Foggy Bottom is thought to have received its name due to the same combination of low-lying marshy areas near a river where fog could sink and stay a while. Other anecdotal stories attribute the DC Foggy Bottom nickname to a collection of industrial smoke from a nearby gas plant.

I find it hard to believe there are foggier places in America, but the geography map from City University in New York shows us that we are in the second foggiest zone.[19]

When you live on an island dominated by fog you come up with quirky animal husbandry related terms like *Sheep Storms*: "Sheep storms were what the islanders called those periods of intense fogginess which often preside over the moors in late June and early July. In the 1800s, the islanders knew these days were coming, and they also knew that after several days the fog would drench the thick coats of the many sheep grazing on the moors (in pasturing sections called the Sheep Commons). They also knew that after the fog inevitably came hot, sunny days during which the coats would dry. That was the time for shearing, which meant the famous Sheep-Shearing festivals, with fun and games and food for one and all. "[20]

Fog has caused many a tragedy at sea in the treacherous waters surrounding Nantucket. One of the most traumatic accidents was the sinking of the Andréa Doria on a calm summer night about 60 nautical miles southeast of

Nantucket. On the night of July 25, 1956, in heavy fog, the Italian ocean liner SS Andréa Doria, inbound to New York, was rammed by the SS Stockholm bound for Europe. Fifty-one passengers died at the point of contact for the Andréa Doria and five crewman died on the Stockholm. Although the Andréa Doria was equipped with state-of-the-art radar, a combination of errors on the parts of both captains doomed the ships.[21] Because the Andréa Doria started listing to starboard almost immediately after the catastrophic collision, half of its lifeboats could not be accessed. Calamai's distress call was radioed to other ships. A freighter, several naval vessels and the luxury liner Ile De France quickly reached the ship in the heavily trafficked area. The Stockholm, not in any danger of sinking, helped with the rescue. The rescue operation continued through the night, saving 1,660 passengers and crew. It's considered the greatest peacetime rescue operation in maritime history. The shipwreck is in 250 feet of deep cold and treacherous waters. Sixteen divers have met their demise diving on this site to recover artifacts from the beautiful ship. The difficulty of the diving conditions and the silty "blizzards" of disturbed sediment have earned the shipwreck the name of the "Mt. Everest of dive sites." [22]

You may have noticed the beautiful Lightship Nantucket LV-112 in the harbor this week; these floating lighthouses provided critical navigational support for a coastline dominated by fog and danger. I found this quote by Bernard Webber, former LV-112 crew member (1958-1960) especially vivid; "The only terror I felt was when on Nantucket Station in rough foggy weather a Radar Target would be observed headed directly towards the Lightship; as it got close you could hear its engines and soon out of the fog so close you could spit on it would come one of the great liners sailing the seas at the time like the S.S. United States or S.S. France etc."[23]

One last story about fog on small islands. Perhaps you have heard of the Yoho of Tuckernuck? Parents would tell their children harrowing tales of a gargoyle-like creature with gleaming red eyes and big bat wings that lurked in the mists of the moors and seashore. Why did parents create these tales? One theory claims it was to keep their children from wandering off in the night to get lost in the fog and swept out to sea by an errant wave or from falling off a cliff.[24]

On that sobering note, I'll end this essay with a random safety reminder. Some folks might not remember that you should always use your low beams and not your "brights" or high beams when driving in fog. Fog droplets form multiple reflective surfaces that disperse high beams and reduce visibility. The next time you are navigating our winding roads in the fog, be safe, watch out for deer, and enjoy the Grey Lady.

What Lies Beneath: Mapping the Ocean Floor
(September 12, 2013)

Photo by Theresa L. Smith, research geologist, Marine Geology Department, Provincetown Center for Coastal Studies

A scientist brought a new "toy" to the UMass Boston Nantucket Field Station and boy what a toy it is! In late August, Dr. Mark Borelli of the Provincetown Center for Coastal Studies, in collaboration with the Nantucket Field Station, brought his 22-foot Parker type boat outfitted with a shallow water multi-beam high resolution acoustic device down to Nantucket to map the harbor and sea floor off the east coast of Nantucket. Each year he'll come down and map more of our seafloor as we gather data to better understand the shifting sands and eelgrass beds around the island.

Mark is investigating the bathymetry and composition of the seafloor.[25] Bathymetry is the study of the underwater depth of lakes or oceans. In other words, bathymetry is the underwater equivalent of topography. The name comes from Greek words βαθύς (bathus), for "deep" and μέτρον (metron), for "measure."[26] Bathymetric (or hydrographic) charts are typically produced to support the safety of surface or sub-surface navigation, and usually show seafloor relief or terrain as contour lines (called depth contours or isobaths) with selected depths (also called soundings) shown on charts in order to provide surface navigational information. No one wants to find a large rock or wreck the hard way! The term "sounding" derives from the Old English word "sund," meaning "swimming, water, sea;" it is not related to the word "sound" in the sense of noise or tones.[27] With the use of acoustic technology we can arrive at a satisfying return to its homonym.

Originally, bathymetry involved the measurement of ocean depth through depth sounding. Early techniques used pre-measured heavy rope or cable lowered over a ship's side. This technique measures the depth only a singular point at a time and is almost comically inefficient. It is also subject to movements of the ship as it is tossed about in the waves and currents moving the line out of true. These complications change the angle of the weighted line and must be accounted for to get a true depth. I find when I used a weighted Secchi disk in the jetties channel with an outgoing ebb tide that can be several knots in speed that even with a relatively heavy lead weight on the bottom it still can fly like a sail and refuse to sink straight down. A Secchi disk is a contrasting black and white disk invented in 1856 by Pietro Angelo Secchi to measure the transparency of water and was modified to the current black and white wedges used today by George Whipple (cofounder of the Harvard School of Public Health) in the late 1880s.[28] Mine are so old that I am pretty sure I have Whipple's original secchi disk. It is a testament to the endurance of this method that we still tie a heavy object onto a line marked at intervals to determine the bottom. Today most mariners use some type of "fish finder" or acoustic device to see differences in density whether from the bottom or schools of fish.

Traditional terms for soundings are a source for common expressions in the English language, notably "deep six" (a sounding of six fathoms).[29] The distance of a fathom which is approximately six feet was determined as the wingspan or distance between the average man's outstretch arms (also roughly proportional to height) used to measure each wet section of line coming up over the side of the boat as it was pulled in. On the Mississippi River in the 1850s, the leadsmen (who measure water depth with a lead line)[30] also used old-fashioned words for some of the numbers; for example, instead of "two" they would say "twain". When the depth was two fathoms, they would call "by the mark twain!" Many people think that the American writer Mark Twain, a former river pilot, took his pen name from this cry. The real story is more interesting. Gregory Crouch provides the details:

Clemens had become widely known in Virginia City — if not necessarily widely liked — by the time the pseudonym Mark Twain first appeared in the Enterprise on February 3, 1863. A decade later, Clemens claimed he'd appropriated his by-then-famous nom de plume from a staid Mississippi riverboat captain. However, according to more convincing Virginia City legend, Clemens acquired the nickname before it appeared in print, derived from his habit of striding into the Old Corner Saloon and calling out to the barkeep to "Mark Twain!" a phrase Mississippi river boatmen sang out with their craft in two fathoms of water, but that in Virginia City meant bring two blasts of whisky to Sam Clemens and make two chalk marks against his account on the back wall of the saloon.[31]

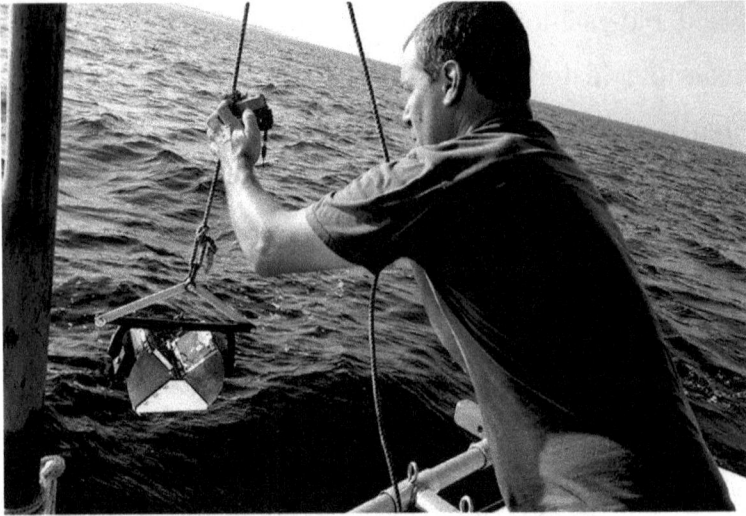

Dr. Mark Borelli deploying a ponar grab to obtain a sediment sample. Photo by Theresa L. Smith.

Let's return to our discussion of the superior technology used today, which can find and create an image of a beer can discarded from a boat. The instrument used was originally leased by Dr. Mark Borelli to complete a project for the Massachusetts Department of Environmental Protection (DEP) Coastal Zone Management (CZM) to investigate sediment transport and coastal features from Race Point in Provincetown to Wellfleet. I am by training a chemical oceanographer and have spent many years taking offshore sediment samples to determine sediment ages using radiochemistry (study of matter using radioisotopes) so I was keenly interested in the high resolution side scan mapping technique. The images I saw of the Cape nearshore bathymetry were vivid and very informative. I spoke with Mark earlier this year and we came up with a plan to take his show on the road down to Nantucket. By helping to purchase the equipment, we were able to make sure Mark had it on hand all the time and we could commission acoustic mapping of Nantucket's coastal waters. The equipment used is a state-of-the-art interferometric sonar system that collects both coincident (matching point for point) bathymetry and acoustic backscatter imagery.[32] Backscatter is the retroflection of waves, particles, or signals back to the direction from which they came. These two data sets collectively provide more information when overlaid on each other to yield a very detailed image of the seafloor, including eelgrass beds, wrecks, rocks and other items of interest. The acoustic backscatter data will ensonify (great word!) or 'take a picture' of the seafloor using sound. Ensonify (also spelled insonify) is an acoustic term which means "to fill the ocean or any fluid medium with acoustic waves to locate or image objects within it (the phenomenon of the acoustic waves bouncing off physical objects is called acoustic radiation)"[33] This technique along with the collection of sediment samples and underwater imagery and/or video can provide detailed information about the surficial seafloor geology including grain size, composition, bedforms (rivulets, wave forms, mounds and sandbars) as well as submerged aquatic vegetation like eelgrass and other biota.[34]

The boat Mark uses has two different sonar systems, one on a boom that is lowered in front of the boat and one that is towed behind. The distance between these two objects is known. Phase Differencing Bathymetric Sonar (PDBS) systems, also referred to as interferometric sonar, use the measurement of phase at each of several receive elements (in the front and back of the boat there are multiple receivers) to determine the angle from which the acoustic return originates. What does this mean in plain English? Systems equipped with this have several receivers just a few inches apart. An acoustic signal is sent from a pinger which bounces the sound down to the sea floor a little like we do when we yell across a canyon and listen for an echo. As that echo bounces off object, each receiver "hears it. Since we know the distance, each receiver is from the other and we can hear the sounds return on the front and back of the boat, we can calculate the angles of return of the objects bouncing the sound and get a very accurate picture of the size and shape of the object with centimeter accuracy. This also allows for accurate imaging in shallow watsers.[35] In his first surveys, Mark found large (up to six feet in diameter) boulders off 'Sconset as he did his surveys from Sesachacha to Low Beach and many mooring blocks and a sunken boat or two scattered around the harbor. Over the next few years, we will be mapping the south, west, and north sides of the island. We will also be sharing his information with Charlie Costello who maps the eelgrass beds around the state for Massachusetts' Department of Environmental Protection.[36]

This equipment is efficient, reliable, and can be deployed in very shallow water. It also collects data in a very wide swath (10:1 versus the 3:1 swath of typical multibeam bathymetry system) which means it can image more of the ocean floor, collecting data faster and allowing for more research.

Mark also uses this equipment to monitor how sand placed in nearshore coastal areas moves around and can be used to heal or augment beaches and sandbars destroyed in ocean storms. He collects sediment samples using a Ponar grab[37] and attaches a "Go-Pro" underwater camera to record the sediment grab and evaluate the sea bottom. An innovative approach that he used in a project in collaboration with Dr. Anamaria Frankic, Director of the Green Harbors Project, involved using an acoustic system to map oyster beds in rehabilitated areas to make sure the oysters have survived their implantation after restoration efforts.[38]

Salt Spray - Like a giant margarita
(September 15, 2011)

Here is an island riddle: We live with it every day. Almost every plant on island has some tolerance to it. Many of our plants can't live without it. Other plants can die if they encounter large amounts of this mystery substance. Give up? What we are talking about today is salt. Much of our island plant life has some salinity or salt tolerance but even those defenses can be strained by excess salt spray which can negatively affect plants after a "dry" tropical storm. Fortunately, here on Nantucket, Hurricane Irene was not the menace originally predicted when it swung by in late August 2011. Folks on the mainland (as islanders often call it, "America") from Vermont to Virginia bore the brunt of the storm's wrath delivered by extremely high rainfall and driving winds. But the island did take a beating to some of its plant life in the form of salt spray damage. It is hard to imagine how the salt spray that is normally a gentle, constant companion on island can quickly accumulate in leaf burning amounts. The majority of plants on Nantucket are relatively well adapted to salt spray. They have little choice on a 50-square-mile island 26 miles from the mainland. The most noticeable and almost shocking consequence from the latest hurricane/tropical storm was the preponderance of damage to plants on the south shore.

You can track the strength and direction of the prevailing winds by walking in a straight line from Cisco or Smooth Hummocks to the north.

Megan Griffiths of Tufts University found that although it was anecdotally understood that salt spray damage from a tropical cyclonic storms could be pretty severe, especially if little rainfall accompanied it, few scientists had gone to the trouble to quantify the effect, much less evaluate the time it takes for plants to rebound. Hurricanes like Irene are a classic example (when it came into range of Nantucket) of an encounter with the "dry side" of a tropical storm which, in this case, is the west side of a north-tracking storm.[39] Without large amounts of rainfall, leaves do not have an opportunity to shed the salt before it begins to affect the tissue, causing browning, "burning", and eventually necrosis (cell or tissue death). Griffith conducted a series of measurements on common coastal heathland species including northern bayberry (*Morella pensylvanica*) before and after Tropical Storm Floyd. Floyd became one of the most powerful hurricanes in decades as it moved across the Atlantic, strengthening to hurricane force on September 10, 1999 230 miles (370 km) east-northeast of the Lesser Antilles.[40] By the time in reached the upper east coast it had been downgraded to a tropical storm. Griffith found that plants not only suffer from the physical damage from the winds but also from storm surge flooding and salt spray. She reported that the most salt related damage occurred during high wind events with low rainfall and that leaves could accumulate twice as much salt in one event than they encountered all season. Perhaps not surprisingly, the most susceptible plants were those further from shore who had sequestered themselves in a zone where salt spray is less common or in lower concentrations. Like fair-skinned city dwellers turning red at the seashore, these more delicate inland plants were at higher risk to leaf browning and loss from the salt spray.[41]

Salt from sea spray acts very differently depending on the time of year it occurs and whether it is concentrated in soil or on the leaves of plants. In spring, plants are much more vulnerable to salt spray. In fall, many trees have survived significant storms because they were in the process of dropping their leaves and typically the evergreens might fare worse during those events. Rainfall is key. If the soil is not saturated by salt water or it is pre-saturated with fresh water from rain or rinsed out sufficiently afterwards, most plants can bounce back. Many gardeners are able to treat salt-soaked soil with gypsum or limestone to counteract the salt. Fertilizers are a bad thing to add because most of those contain salt.

Many people are very aware of the harrowing record of lives lost and buildings and landscapes destroyed by the 1938 east coast hurricane better known as the Long Island

Express. What many people don't consider is how salt spray can weaken plants near the ocean and reduce the resilience of coastal areas to erosion simultaneously disrupting natural biodiversity. From one account:

"Salt spray from wind-blown sea water and mixed rainwater also had the effect of browning trees that did survive. Weeks later these trees were dead. One can still find many downed trees throughout eastern Long Island's forest that are a direct result of this massive hurricane. More recently, a study of the Buzzards Bay coastal region revealed that 50% of the salt-sensitive White Pines were killed by salt spray from Hurricane Bob in 1991. In 1938, Long Island salt marshes were inundated with tons of overwash (sand brought over the dune into marshland areas) that prevented the marsh grasses from growing back. This effectively decreased the area of the salt marsh for years."[42]

Botanists at the Arnold Arboretum at Harvard University painted a much rosier picture of the recovery rate of plants after a major storm. The authors systematically observed various plants in the area to evaluate how quickly the plants would bounce back after exposure to hurricane force winds and salt in Woods Hole, Falmouth, and the north shore of Massachusetts and Newport, Rhode Island. I especially like this extremely optimistic (and misguided quote) "Fortunately, with hurricanes in the east spaced 100 years apart, it is not necessary that the fear of another in the immediate future should govern present seashore planting".[43] That statement is a knee slapper. One of the many plants that survived relatively unscathed was the Japanese pitch pines and the currently blooming groundsel tree (Baccharis halimifolia). Overall, the authors were extremely impressed by the ability of many plants (each divided into hardiness after salt exposure) to recover.[44] Botanists associated with Arnold Arboretum did the same type of evaluation after hurricane Donna. They found that "good soil" (more organic matter than sand) and warmer winter temperatures helped plants to survive a "good salt dousing".[45]

One last, somewhat random silver lining came about because of the Long Island Express. One of the few saving graces for storms like this (and hopefully some solace for those suffering from Irene) is that the repair and recovery efforts can create jobs. As an example: "One positive economic outcome of the 1938 Hurricane was that it effectively ended the unemployment experienced near the end of The Great Depression. At that time most people were out of work and would gladly work for the standard wage of $2 per day. Because so much damage had occurred to homes and buildings and so many trees were blocking roadways, thousands of people flocked to Long Island in search of clean-up work and repair. In fact, more than 2,700 men were brought into New York and New England by Bell Systems just to repair the downed phone lines."[46]

The Halophytes: Salt Lovin' vegetation
(September 15, 2011)

In contrast to our trees and shrubs, which tolerate salt spray only to a certain extent, there are many coastal plants that only thrive in salty habitats, where their special adaptations give them an edge. Therefore, we have salt lovers and salt haters, with or without a margarita's involvement. Salt loving organisms are called halophiles from the Greek words "hals" whose masculine meaning is "salt" and feminine meaning is "sea" and "philos" for "love."[47] Most salt marsh plants are easily able to survive hurricanes IF they are not physically pulled out by large waves or tidal surges. Smooth cordgrass (*Spartina alterniflora*), dominates the regularly flooded low marsh. Smooth cordgrass is the most abundant salt marsh plant on Nantucket and is responsible for much of a marsh's productivity. Smooth cordgrass' successful adaptations enable it to live where few other plants could survive. It has narrow, tough blades and special epidermal salt glands that secrete excess salt, enabling it to withstand daily exposure to saltwater and higher temperatures. Only a few animals like snow geese eat this plant, but many animals and plants live on it or on the marsh surface protected by its roots and stalks. Smooth cordgrass stalks are thick, very tough, and well anchored by a root system.[48]

From a distance, the low marsh appears to be uniform; however, there are two forms of smooth cordgrass. A tall form that can reach heights of six feet or more grows along creek banks and in old mosquito ditches and clings to the wettest, most often submerged parts of the salt marsh. A short form of smooth cordgrass occurs in slightly drier interior parts of the low marsh and ranges from two to three feet in height. In contrast to the low marsh which has one major species of plant, the high marsh contains a mixture of several species including black rush (*Juncus gerardii*), salt hay aka salt meadow cordgrass (*Spartina patens*), some short-form smooth cordgrass, and glassworts (*Salicornia spp.*) which are very tasty in salads. This high marsh area grades into a marsh-upland border which is a transitional zone between the salt marsh and the maritime shrub community that consist of groundsel tree (*Baccharis halimifolia*), bayberry, poison ivy, seaside lavender, and a shrub known by a few names like high tide bush and marsh elder, specifically the southern maritime marsh-elder (*Iva frutescens*).[49]

Salt marshes are transitional areas between land and water, occurring along the intertidal shore of estuaries and sounds where salinity (salt content) ranges from near ocean strength (33-35 parts per thousand or ppt) to near fresh (zero ppt) in freshwater marshes. As we mentioned above, plants grow in different tidal zones based on their salt tolerance, with the highly-adapted grasses of the smooth cordgrass and salt meadow cordgrass growing directly in the water and making up the majority of the marsh, marsh elder and groundsel shrubs in an intermediate zone, and cattails (*Typha spp.*) and the invasive common reeds (*Phragmites Australis*) existing in the very low salinity upland part of the marsh where fresh water flows in. Nutrients brought in by rainwater from the uplands combined with an abundance of decomposing organic matter makes the salt marsh one of the most productive habitats on earth, and an important breeding ground and nursery for saltwater fish, shrimp, crabs, and shellfish including mussels and snails.

A large amount of biomass is created and consumed in the average salt marsh. There hasn't been a movie like "The Lion King" written about the circle of life in the salt marsh but it is occurring every day in a life and death struggle both above and under water. In spring and summer, marshes are lush and green, highly productive and growing in height. In late fall and winter, the green smooth cordgrass begins to turn brown as leaves die and decomposition begins. Water, waves, wind, and storms dislodge and break up decaying leaves, and transport them to mud flats and other locations around the marsh. This dead plant matter, or detritus, forms an attachment site for microscopic organisms such as bacteria, fungi and small algae. These organisms colonize the broken bits of plant material and break down portions of the detritus that are not digestible by animals. This occurs best in a reducing environment where oxygen concentrations are low, and this gives salt marshes their sulfury rotten egg smell (from hydrogen sulfide which is released from the muck).

One of my favorite plants on Nantucket is just starting to bloom in mid-September, and this is a plant that can withstand a lot of saltwater stress. It also is the epitome of "truth in advertising." I am referring to the groundsel tree or groundsel bush (*Baccharis halimifolia*) sometimes called "high tide plant" or salt marsh elder or sea myrtle. I am surprised it hasn't changed its name to a symbol ala the late great Prince (R.I.P).

Groundsel bush/tree is the only native eastern species of the aster family reaching tree size. Baccharis is the ancient Greek name (derived from the Roman god of wine and intoxication, Bacchus) of a plant with fragrant roots. The Latin species name means "with the leaves of Halimus", an old name for Saltbush, an unrelated shrub.[50] On February 16, 2018, the common name of this species was changed in the United Stated Department of Agriculture Fire Effects Information System (FEIS) website from groundsel-tree to eastern baccharis which is honestly a relief.[51] In fact, if you look up groundsel on the USDA plants database website you will find 42 plants, none of which are eastern baccharis!

There are many different "high tide" shrubs which demarcate the average height of the wintertime high tide. If you look out over Folgers' Marsh, you'll see *B. halimifolia* and the very similar marsh elder (*Iva Frutescens*) arrayed around the marsh at the highest wrack line. In order to tell them apart, eastern baccharis (aka groundsel tree) has alternate leaves and marsh elder has opposite leaves which are a bit pointy-er and they have very different flowers. Dr. Rick Kessler and his students at the University of Massachusetts Boston has been studying the genetic diversity of these plants at the Nantucket Field Station for several years. In fact, Dr. Kesseli's research necessitated me bringing male and female full-sized plants encased in trash bags (to keep them from "mingling"- avert your eyes, children) in buckets onto the Steamship ferry to transport them to campus. Good times.

Eastern baccharis' numerous branches emerge from short trunks which are covered densely with branchlets. The 6-12-foot deciduous shrub bears gray-green, somewhat lobed, oval leaves which are semi-persistent in the North (evergreen-ish; unlike my house plants). White to green flowers occur in small, dense, terminal clusters. Probably the most significant feature which you'll be able to see all over Nantucket is the silvery, plume-like achenes (simple dry fruit) which appear in the fall on female plants which to me look like silvery paintbrushes.

These plants exhibit dioecy (Greek: διοικία "two households"; adjective form: dioecious), which means the male and female plants are segregated.[52] Male plants generally have longer shoots, more tender leaves, grow faster, and flower and senesce (grow old, age) earlier than female plants. Their reproductive organs look like little soccer balls with points on them. When they release their pollen into the air the entire border of the salt marsh is covered with golden dust. Before I "met" this plant, I had not thought about male and female plants being separate entities. Every late summer/early fall, as I walk around the marsh, I marvel at their resilience.

Impacting the Globe – The Anthropocene
(August 6, 2015)

In August of 2015, I was fortunate to be hosting journalist Andy Revkin for an on-island event to support the Organization of Biological Field Stations (OBFS)[53] in my role as the President of OBFS. Andy was the science and environmental beat writer for the New York Times for 14 years and wrote an excellent science blog for the New York Times called Dot Earth for nine years.[54] He has written extensively on climate related issues including the evolution of the Arctic and the crisis of communication that stymies many scientists as they try to describe their methods and conclusions to a public unsure of the scientific process. He is the founding director of the new Initiative on Communication and Sustainability at Columbia University's Earth Institute. Andy coined the term "Anthrocene" in his book *Global Warming: Understanding the Forecast* (1992), in which he wrote, "we are entering an age that might someday be referred to as, say, the Anthrocene. After all, it is a geological age of our own making."[55] The name evolved into the term scientists use now of Anthropocene ("anthropo" from the Greek words, "anthropos" (ἄνθρωπος) for "human" and "cene" for "new") to describe the current geologic age as one in which human impacts are causing ecosystem wide change and have become the dominant driver of many earth processes.[56]

Andy reminded me of a group that came here for the 2010 "Living on the Edge" Coastal Communities conference called *Atlantic Rising*. *Atlantic Rising* was a two-year charitable project created by a group of three young Brits (Tim Bromfield, Will Lorimer, and Lynn Morris) who traveled the world showing people what a one-meter (three foot) sea level rise would look like on the ground.[57] They took their blog, videos, and other material and morphed it into educational resources for students aged 12-15. I spoke with the group, and I was really impressed by their mission and their ability to viscerally show an average person what sea level rise would look like. They filmed a video with Andy with high school students on Nantucket working with local surveyor Art Gasbarro to measure and mark the one-meter level at Brant Point using yellow caution tape.[58]

I don't think we necessarily need caution tape to remember that Nantucket, like many other low-lying islands, is extremely vulnerable to sea level rise. The recent winter storms over the past three years have drilled into us the need to plan for sea level rise and to prepare for these storms in intelligent ways. The Town of Nantucket developed a Coastal Management Plan to address town-owned land in harm's way from storm surge and erosion that was completed in 2014.[59]

The Cape and the Islands were formed during the Wisconsin glaciation period when a lobe of the Laurentide Ice Sheet (named the Laurentide after the Laurentian region of Canada where it first formed) pushed rocks and debris south as it flowed over the area. In fact, the ice sheet moved back and forth a few times (advanced and retreated), but essentially it stopped right along the area on Nantucket where we have a few hills on the northern side. This line is called a terminal moraine, and there are adjacent ones formed from other glacial lobes forming the backbone of hills that can be seen along Martha's Vineyard and the Cape.

The southern side is mainly outwash plain, or bits of sediment that melted out and stayed there as the ice retreated north as the earth warmed up. The islands became islands as the sea level rose around 10,000-12,000 years ago. Before that, man or beast could have walked to Nantucket. The wind and waves started moving the sand around, redistributing it in storms and depositing it just like a pile of sand in a bucket with some water would do if it were swirled around.

The sea level rise that created Nantucket's watery moat that separates us from the mainland ("America" as locals call it) is continuing at a faster pace. Worldwide, the conservative midrange estimate predicts oceans will rise 18 inches by 2100 [Author's note, this is a 2015 prediction, in 2022 that number is now 1 meter!]. As you may have noticed, Nantucket is not very high, Altar Rock is only 100 feet above sea level (fourth highest point) and the highest point, Sankaty Head, just a little south of Sankaty Light clocks in at a dizzying 111 feet. If sea level rise stayed constant at a rate of almost a foot over the last 100 years, we would have plenty of time to enjoy the island, but it is accelerating, and the loss of marshes and low-lying land means storm surges are going to move inland. The New England area has experienced accelerated sea level rise over the last 500 years from 1 mm per year to 2.5 mm/year (in 1990) to close to 3.0 mm/year now in 2019. The most accurate recent estimates for Nantucket can be found at the National Oceanic and Atmospheric Administration (NOAA) sea level website. You can see the winners and losers across the nation quite quickly and there are far more losers than winners. Right over Nantucket a yellow arrow pointing up indicates that for Nantucket "The mean sea level trend is 3.58 mm/year with a 95% confidence interval of +/- 0.39 mm/year based on monthly mean sea level data from 1965 to 2014 which is equivalent to a change of 1.17 feet in 100 years."[60]

Massachusetts includes 1500 miles of tidal shoreline, 78 coastal communities, 681 barrier beaches and 36,000 people live within 500 feet of a coastal shore. Approximately 72% of the Massachusetts shore is exhibiting a long-term erosional trend and this trend has accelerated since 1950. Nantucket has the highest erosion rates in the state, with parts of the southern side losing 12 feet per year on average.[61] Storm generated erosion ranges over periods of hours (tropical cyclones) to several days (northeasters). Although the storm events are short-lived, the resulting erosion can be equivalent to decades of long-term erosion. The actual quantity of sediment eroded from the coast is a function of storm tide elevation relative to land elevation, the duration of the storm and the characteristics of the storm waves. During severe coastal storms, it is not uncommon for the entire berm (dry beach above the normal

53

high-water line) and part of the dune to be removed from the beach. The amount of erosion is also dependent on the pre-storm width and elevation of the beach. Repeated small storms can do a lot of damage because the beach is more vulnerable to sand loss. In fact, the cumulative effects of two closely spaced minor storms can often exceed the impact of one severe storm as we saw this past spring with the nor'easters "Nemo" in February and "Saturn" in March occurring not all that long after Superstorm/Hurricane Sandy and doing more damage than Sandy to many parts of the island. Sandy was the second most costly hurricane in the U.S (68 billion dollars) only surpassed by Hurricane Katrina. Therefore, although hurricane erosion can be serious and dramatic, in the long run, it is the northeast storms that do the most damage.[62]

In January, winter storm Juno once again battered the island, taking away a large chunk of the barrier beach here at Folger's marsh, driving sand all the way to Polpis Road and marking a low lying near shore shack that use to be my home and office with a mud line not far from the one deposited by Hurricane Bob (August 1991) and others lines left by hurricanes and storms like Sandy, Saturn and Nemo. These two, three, and four-foot mud and water lines on the Field Station "beach house" remind me of the heights of children on a door frame. Juno also ripped huge refrigerator sized chunks of marsh vegetation and ribbed mussels out of the marsh and hurled them like bowling balls onto the back marsh. By far the damage to the Nantucket Field Station land was some of the most extreme I have seen in the past 12 years. Boston recorded a record-breaking 22.1 in (56 cm) of snow, its largest January storm total accumulation and its sixth largest storm total accumulation on record at the time [63] and we had a long period with no electricity on island.

The fact that our sand moves around helps us recover from storms like these as the sand moves offshore to form sandbars and shoals. Nantucket's sand acts as an amoeba or a moving shield, naturally migrating offshore and back onshore as it is driven by waves, wind, and tides. Nantucket as an island is relatively nimble and able to adapt to change. But within 200-300 years, much of the island will be flooded and only the higher elevations will still be sticking up out of the ocean. I give a talk about three or four times a year on the effect of climate change on Nantucket. One of the tools I use in the talk is an online topographic based sea level inundation website called Firetree which is relatively accurate and based on NASA data.[64] Firetree and other FEMA based software and online data mappers like NOAA's Digital Coast[65] software by the Center for Coastal Management allows you to zoom into your street or neighborhood and see what modelers are predicting for sea level rise and storm inundation. These tools simply take topography and overlay one, two, or three plus feet of sea level rise on top. The Massachusetts Sea level rise and coastal flooding viewers allows one to layer FEMA flood maps and hurricane storm surge to a one- or two-feet sea level rise.[66] It is a pretty sobering wakeup call when you input a one-meter rise in sea level and see most of Coatue, Madaket, and parts of downtown, Brant Point, and the marsh edges at the field station under water. So far, island marshes have been relatively stable and have built up to keep pace with the rising sea level, but higher astronomical tides, more intensive storms which accelerate erosion, and shoreline development projects which divert the sediment flowing to and along our shores can hurt the ability of the island to adapt to sea level rise. For the twelve years I lived on Nantucket I had a front row seat during each nor'easter to evaluate coastal erosion.

Changing, perhaps abruptly, to another strange and recent story about the effects of sea level rise and warming temperatures: in 2015 a story started circulating on the web describing how climate change and warming seas have led to dolphins moving further north up into the Arctic area as sea ice melts. Polar bears are quickly adapting to this new food source. In one instance a bear was observed killing and eating one dolphin and stashing the other one under the ice to eat later! That is an impressive feat that has been seen a few times before when hapless whales get trapped by refreezing ice and devoured by hungry polar bears.

In a research paper published in June 2015: "Polar bears have been spotted eating dolphins in the Arctic for the first time ever by Norwegian scientists, who believe global warming could have caused the bears to expand their diet. Polar bears feed mainly on seals but Jon Aars at the Norwegian Polar Institute has photographed dolphins being devoured by a hungry bear. 'It is likely that new species are appearing in the diet of polar bears due to climate change because new species are finding their way north' he told the AFP (Agence France-Presse)".[67] As our oceans warm and become more acidic and areas formerly ice covered year round are exposed, stories like this may become commonplace. How humans and other animals adapt to these changing conditions as the Anthropocene progresses will hopefully not be a swan song but a song of resilience.

Is Nantucket Sinking?
(May 30, 2012)

In May of 2012 in a Q&A forum of www.Nantucket.net a question was submitted by a teacher named Vienna. She wrote: "Is Nantucket sinking? My class is doing a state report and my students need to know if Nantucket is sinking. Google is not helping." Dear Vienna: The short answer is: No, Nantucket is not sinking. The longer answer is, although Nantucket is not sinking, sea level is rising, and we'll eventually be under water either way.

I then went on to elaborate with a shorter version of the explanation below. Google was a little more helpful to me, and I was able to turn to several books that explain the processes involved with land sinking or rising. Hydrologists call the occurrence of land sinking subsidence. Subsidence is the motion of a surface (usually, the Earth's surface) as it shifts downward relative to a datum (fixed starting point) such as sea-level. In order to measure land sinking or sea level rising, you must have a stable constant known height of land to measure the change in elevation. Determining that point when everything is moving, even extremely slowly and almost imperceptibly, is not as easy as you think. Scientists and the US government began releasing much more accurate estimates of mean sea level (height between low and high tide averaging out tides and storms) in 2015.

Ground subsidence is of concern to geologists, geotechnical engineers and surveyors, and to anyone living within 20 feet of sea level or the unlucky few who have cars that have fallen into sinkholes. Common causes of land subsidence from human activity are pumping water, oil, and gas from underground reservoirs; dissolution of limestone aquifers (sinkholes); collapse of underground mines;

drainage of organic soils; and initial wetting of dry soils (hydrocompaction). Land subsidence occurs in nearly every state of the United States. Massachusetts is not considered to be significantly impacted by subsidence across the entire state but areas like Boston are suffering because it is composed of fill on marshland. The mid-Atlantic region as a whole is subsiding as a result of the removal of a large layer of ice after the last "mini" ice age for the east coast around 10,000-11,000 years ago.[68] The news is not good for parts of Louisiana which has some of the highest subsidence rates in the world.[69]

The opposite of subsidence is uplift, which results in an increase in elevation. In fact, most of New England at one time was undergoing isostasy which is one of my favorite scientific terms. Sounds fun, doesn't it? This is the opposite of subsidence and causes uplift of the land, which naturally results in an increase in elevation. Glacial isostasy is the process of lithospheric depression beneath the weight of an ice sheet and subsequent rebound when the ice mass is reduced or removed. When we were covered in ice, it pushed the land down with its weight and when that ice retreated, the land started bouncing back, slowly rising over thousands of years.

Glacial eustasy, on the other hand, refers to worldwide changes in sea level because of the changing volume of glacier ice on land. Both glacial isostasy and eustasy are related to the volume—thickness and areal coverage— of ice sheets. You may remember from school that our continents float around on top of a plastic fluid asthenosphere just like ice cubes in a drink or logs in a river. The asthenosphere is a portion of the upper layer just below the lithosphere that is involved in plate tectonic movement and isostatic adjustments. If mass is added to a local area of the crust (for instance by deposition), the crust subsides to compensate and maintain isostatic balance. If Godzilla pushes down on a large ice sheet on one end, the other end lifts out of the water, that also goes for ocean liners.

Although the state as a whole and Nantucket itself is not sinking, there are some areas of New England's shoreline that are sinking due to subsidence. Scientists have determined that the area around Boston (which contains lots of fill) may settle as much as six inches in 100 years; which is quite a bit. The Boston area has filled in close to two-thirds of its wetlands, which would have protected it as sea level rose.

"British colonists founded Boston in 1630 next to a freshwater spring on the heavily forested Shawmut Peninsula. By the 1800s, the trees had been replaced by a bustling trading port. As the population grew, industrious residents began filling in tidal flats and marshland with rocks, dirt and trash to create more buildable space. By the early 1900s, the city had tripled in geographic land area. The

South End, Charlestown, East Boston, Back Bay and downtown neighborhoods, including attractions like historic Faneuil Hall and the New England Aquarium, are all built on landfill. Even Logan International Airport is built atop a filled-in tidal flat that was once five islands."[70]

And in the past few years, the prognosis for Boston's sea level rise has risen (pun intended) to indicate a 2 to 3.3 feet (0.6 to 1 meter) rise due to climate change and natural land subsidence.[71]

To make matters worse, we all know that independently from subsidence is sea level rise due to our warming climate. Worldwide, the conservative low emissions estimate predicts oceans will rise one foot by 2100 over 2000 levels (2019 estimate).[72] Louisiana faces a far more alarming forecast. Here, those same models predict that relative to the land, water levels will rise 2 to 6 feet, with the highest rates in the southeastern coast surrounding New Orleans. This is occurring because Louisiana has the highest rates of subsidence, and its salt marshes have been deprived of sediments from rerouting of the Mississippi. Only the Atchafalaya region (where I did some research as a graduate student) is building up sediments naturally to keep pace with sea level rise. When we think about sea level rise, storm surge and coastal inundation, it is important to factor in how subsidence will exacerbate the impact of the rising seas. A person can end up treading water either way.

Shifting Sands, Coastal Beach Processes, and Erosion
(August 11, 2011 and June 20, 2013)

Erosion is the number one hot button topic on Nantucket now and has been for the past hundred or more years. I wrote about erosion and beach processes many times over the years. Nantucket is experiencing between 0.74 feet to 12.0+ feet of erosion each year. Nantucket has some of the highest erosion rates in the state, as the island slowly and inexorably erodes on the eastern and southern shores while currents and tides take that eroded sand and deposit it in shoals to the north and west of the island. Like all islands thrust into the open ocean, yet still close to land, and most barrier islands, Nantucket is slowing migrating north and west to join the "mother ship" of the mainland. Similarly, the islands of Muskeget and Tuckernuck have greatly eroded on the southerly, ocean facing sides and filled in on the slightly more protected northern Nantucket Sound side, essentially migrating slowly northward as sand is reworked and redeposited.

Is this bad or good? What is erosion and why do we need it? Erosion is the wearing away of the earth's surface by any natural process. Around most of the world (think Grand Canyon) the chief agent of erosion is running water; minor agents are glaciers, the wind, and waves breaking against the coast. Everything that we think of when we think of soil or sediment was created by a combination of erosion by wind and water and decomposition by fungi and bacteria. Our beaches would not exist if erosion had not broken down the parent material into fine little particles. Cape Cod and Nantucket are a product of the last continental glaciation, the Wisconsinian glacial stage during which the Laurentide Ice Sheet advanced and retreated. Erosion of glacial landforms like moraines, drumlins, outwash plains, and kames provides the primary source of sand and cobble for Massachusetts' 1,500 miles of beaches, dunes, and barrier beaches.

Over the years I have collaborated with a variety of geologists studying sedimentation in nearshore areas and almost all my research has been on sedimentation rates in harbors and the nearby ocean as deposited by rivers. The entirety of the U.S. except new volcanic land will eventually be worn down and deposited offshore. What is exciting about living on Nantucket is that you can experience this erosion and literally earth moving event in timespans that are accessible to us. It is both unnerving and exhilarating.

At the UMB Nantucket Field Station from 2004 to 2016, we taught visiting and local school kids how to do beach profiling while also conducting long term monitoring with island citizens to document the loss of shoreline in some areas and the build-up in others. This project was started with Jim O'Connell, who at the time was a coastal processes specialist working for the Massachusetts Coastal Zone Management (CZM) division with the Woods Hole Oceanographic Institution Sea Grant Program and Cape Cod Cooperative Extension. The Commonwealth wanted to "ground truth" their aerial surveys and obtain more accurate information on the location of the high tide line to compare to the snapshot they got when flying overhead. In some cases, the on-the-ground measurements indicated that they could be off as much as 150 feet which is a large error when trying to estimate surge impacts compounded by sea level rise. In Massachusetts, relative sea level has been increasing slowly to an average now of 1 foot every 100 years. According to the National Oceanic and Atmospheric Administration (NOAA), The sea level off Massachusetts' coast is up to 8 inches higher than it was in 1950.[73, 74] This increase is mostly due to due to changes in ocean circulation and ice melt.[75] Boston is in even worse shape than other parts of Massachusetts because so much of it (approximately 75%) is built on fill. Its speed of rise has accelerated over the last ten years and it is now rising by about one inch every eight years. That seems like a small number but it adds up to a large storm surge impact especially when you factor in more frequent and stronger storms due to warmer sea surface temperatures.

"Over the last century, sea level in Boston Harbor rose by about 28 centimeters, due to both thermal expansion of seawater as the oceans warm (SN Online: 9/28/18) and the melting of distant ice sheets. Conservative projections for Boston place sea level about 15 centimeters higher by 2030, 33 centimeters higher by 2050 and 149 centimeters (almost five feet!) higher by 2100. In a worst-case scenario, if greenhouse gas emissions continue at the current pace, sea level could rise by as much as three meters by 2100.

Boston is the fifth most vulnerable coastal city to flooding from sea level rise in the United States — after Miami, New York City, New Orleans, and Tampa — and the eighth most vulnerable city in the world, in terms of overall cost of potential damage, according to the World Bank."[76]

But, back to our peaceful beach experiments. Beach profiling is a remarkably simple technique that can be done anywhere with kids as young as eight years of age.[77, 78] Jim is an excellent teacher, great with kids, extremely knowledgeable about beach processes and formation and a joy to work with. Last I heard he was still doing coastal management consultations on Cape Cod.

Enough reminiscing: next, we will cover some beach basics, and no, I am not talking about what type of swimsuit to wear or whether SPF 30 or 60 is better. Different parts of a beach are more or less stable, the back dunes are very stable; the part that encounters ocean storms and wind is less stable. When you measure a beach's topography you are doing what is called a "beach profile" which is not very different from our facial profile from forehead to neck. If the beach were a person lying down perpendicular to the ocean, the forehead would be the back beach and back dunes and the neck would be the foreshore. So, the parts of a beach include, in order of distance from the low tide line away from the water: intertidal area/forebeach, midbeach/berm, and back beach. The following are some definitions of beach terms from a talk I gave at the Nantucket Whaling Museum in 2009:

The "foreshore/forebeach" is the sloping portion of the beach between high and low tide. The "swash zone" is the intertidal area of wet sand where the most recent high and low tide resided. Kids like to play in the swash zone, that is where you will find the mole crabs!

"Wrack line": deposit of vegetation (algae, eelgrass, etc.) and other floating material on the shoreline marking the furthest extent that high water reaches on a beach; wrack lines can usually be seen for the most recent high tide and the highest one of the week and sometimes are even evident for the highest tide of the year.

"Grain size": The grain size is equivalent to the diameter of a sand particle; we use sieves of various hole sizes to sort out the grains of sand on a beach. Low energy beaches have very fine sand, high energy beaches have coarser sand and very high energy beaches are rocky. Wherever you are in the world, you can look down at the beach and get a good idea of the energy of the waves that hits that beach by the size of particles from powder fine sand to giant boulders.

A "berm" is a nearly horizontal rise that looks like a land-based sand bar and is formed when the waves deposit sand on a beach. A storm berm can mark the highest limit of storm waves. Several berms can occur at spring and neap tide levels. The "back beach" or "backshore" is rarely touched by wave action and ends at the edge of the first dune. The "active dune" or "primary dune" is the first dune. "Fixed dunes" or "secondary dunes" can follow, sometimes in great numbers. A swale is the hollow between dunes, often close enough to the water table so that wetland type plants may be present.

"Longshore drift" is what typically controls the movement of sand grains along the beach by waves. Waves that approach the shore at an angle (which is more common than waves approaching head on perpendicularly) rush diagonally up the beach. The water then returns directly down the beach under the force of gravity. Sand grains carried by the rush and backwash of the waves are moved along the beach in a sawtooth fashion. Other grains are carried along just seaward of the beach by the longshore current, which is also generated by the oblique approach (sideways, glancing approach) of the waves. Longshore currents and longshore drift are generally considered to be constructive processes. Unlike storm waves, they are not significant in coastal erosion. They are the continuing processes that nourish the beach and carry sand along the shore. Longshore drift is what reshapes a barrier spit by picking up sand and depositing it at the end of the spit so that the spit grows in length.

In Massachusetts you will find 1500 miles of tidal shoreline, 78 coastal communities, 681 barrier beaches and in 1990, a third of the population lived in a coastal community. Approximately 72% of the Massachusetts shore is exhibiting a long-term crosional trend and this trend has accelerated since 1950.[79]

Storm-generated erosion ranges over periods of hours (tropical cyclones) to several days (northeasters). Although the storm events are short-lived, the resulting erosion can be equivalent to decades of long-term erosion. The actual quantity of sediment eroded from the coast is a function of storm tide elevation relative to land elevation, the duration of the storm and the characteristics of the storm waves. During severe coastal storms, it is not uncommon for the entire berm (dry beach above the normal high-water line) and part of the dune to be removed from the beach. The amount of erosion is also dependent on the pre-storm width and elevation of the beach. Repeated small storms can do a lot of damage because the beach is more vulnerable to sand loss with no ensuing time to build back up its protection. In fact, the cumulative effects of two closely spaced minor storms can often exceed the impact of one severe storm as we saw in 2013 with the nor'easters "Nemo" in February and "Saturn" in March occurring not all that long after Superstorm/Hurricane Sandy and doing more damage than Sandy to many parts of Nantucket. Although hurricane erosion can be serious and dramatic, in the long run, it is the northeast storms that do the most damage.[80]

It is easiest to think of our shifting shoals and tiny island we live on as a system. Most of the sand stays in the system but is transported in storm events offshore leaving much steeper and shorter wintertime beaches. This material sits in a swirling holding pattern and provides protection to ensuing storms by creating underwater sand bars which you can see when you see waves breaking offshore in temporary shallows. During summer's gentler wave climate, with more southerly breezes, the sand moves back into place and the beaches lengthen by many feet, become much flatter and typically a bit lower in elevation. If you only visit Nantucket in the summer and came back in the winter to find your favorite beach "gone" you would faint!

Our beach profiles show this process occurring each year in areas like Codfish Park in 'Sconset, which becomes 50-100 feet narrower in the winter and much steeper. Many beach areas develop a "stair-step effect" as mini escarpments are created when winter storms carve out large chunks of sand. This sand migrates around the island in shoals offshore. Each of these shoals may protect a part of the island for dozens of years, but then it eventually migrates (in the case of the eastern side off Sankaty and near Codfish Park) south; or on the southern shore for instance by Smith's Point and Esther's Island, to the west, exposing areas to erosion that previously were protected. Nantucket has experienced trend reversals, in which sand has built up for many years, in an accretion cycle, then reverses to an eroding cycle to show a deceptively minor average erosion of only a foot or so per year. The time period one examines shoreline changes makes a big difference when trying to figure out if it is eroding or accreting. Over long-time spans, sometimes erosion and accretion are approximately the same, but that is no solace on a decadal or yearly time scale.

For instance, on the southern shore near Surfside, Massachusetts Coastal Zone Management found: "In many cases, short-term shoreline fluctuations can be orders of magnitude greater than the long-term rate of shoreline change. For example, Nantucket's southeast shore has a long-term average shoreline change rate of +0.10 feet per year ("net" accretion of 2.1 feet between 1846-1978), suggesting a relatively stable area. However, between 1846 and 1978 the shoreline accreted 238 feet, then eroded 236 feet. This same phenomenon occurred at Codfish Park on the eastern shore of Nantucket. Unfortunately, many homes were constructed during the accretion phase. Since the trend reversed to erosion beginning in the mid-1950s, many houses have been lost to erosion and storms."[81]

The primary agent of human-induced erosion in a coastal environment is the interruption of sediment sources and longshore sediment transport. Coastal structures interfering with the littoral transport are the most common cause of coastal erosion. Examples include the armoring of sediment sources with seawalls, revetments, and bulkheads, and the interruption of longshore sediment transport by the construction of groins and jetties. Coastal erosion, sand transport, and deposition are the natural processes that are responsible for the Cape as we know it. The cliffs on E/NE side of island erode to form the beach below and to the south and north (nodal point pushes some sand north, some south, node itself moves too). Jetties and groins generally do not stop erosion but interfere with longshore drift and longshore currents to stop the passage of sand along the beach.

When we build groins or jetties, we are diverting some of the longshore drift and transport of sand. Hard structures stop sand from being supplied by a beach front to build the beach directly below or down drift ("down current") from the area. This is called "starving a beach" and you can see in many areas where hard structures like revetments have been built that the beach downdrift disappears unless it is supplemented by additions of sand. If you go to google earth and look at any coastal region, zoom into a hard structure like an old groin or jetty and you will see what I mean. Of course, this was the intent of these original structures. The builders wanted to "steal" some of that sand for themselves and remove it from the system. In some cases, like jetties, the builders wanted to prevent sand from filling in a shipping channel and creating shoals and these structures are augmented by frequent channel dredging.

The field station has been photographically documenting the loss of coastal bluff at Tom Nevers for almost 40 years now as buildings, roads, and tons of bluff material composed of poorly sorted sand, larger gravel and fine clays slough off the beach escarpment face and wash into the nearshore area to be carried along by longshore drift and deposited in sandbars offshore circling the island or on beaches down drift.

Beach Erosion at Cisco

Photo by the author.

Each winter, our beaches become much steeper and shorter. Below is a graph done by Max Duce and Stefan Silverio, former students at the Nantucket New School who worked with me in 2012 and 2013 on a research project to evaluate winter beach dynamics by doing beach profiles at Cisco and 'Sconset. These guys did a great job, especially since my early profiles dating back to 2004 have been buried under 6 feet of sand in the case of Codfish Park because of the drastic shortening of the beach. The best way to explain it is like a six-foot-tall 200-pound person overnight becoming a four-foot-tall 200-pound person, same amount of sand, but a very different profile.

Beach profiling. Photo by the author.

From "Beach Profiling" paper and presentation June 2013 by Max Duce & Stefan Silverio, 8th grade Nantucket New School.

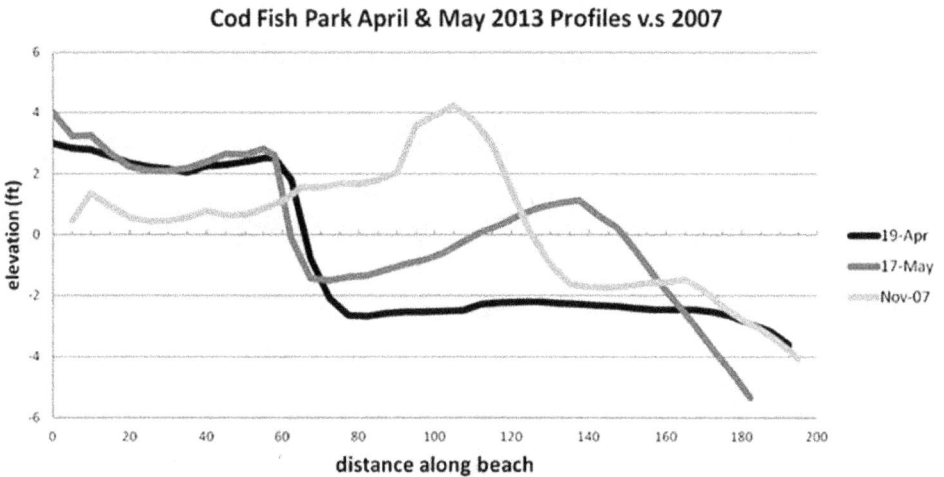

Cod Fish Park April & May 2013 Profiles v.s 2007

According to the Army Corps of Engineers, coastal geologists, and coastal process experts, the most important cause of human-induced erosion is interruption of sediment sources and longshore sediment transport. Examples include the armoring of sediment sources with seawalls, revetments, and bulkheads, and the interruption of longshore sediment transport by the construction of groins and jetties. Coastal erosion, sand transport, and deposition are the natural processes that are responsible for the Cape as we know it. The cliffs on east side of the island off Baxter Road erode to form the beach below and to the south and north emanating from a nodal point that pushes some sand north, some south as winds evolve from winter to summer (and to complicate matters, the node itself moves/migrates a bit). Jetties and groins generally do not stop erosion but interfere with longshore drift and longshore currents to stop the passage of sand along the beach and reduce the natural buildup of sand on a beach. Other types of loss that occur on a beach where revetments have been built include scouring, where banks and beach material is lost on either side of a structure, impoundment loss mentioned above where sand can't be provided any longer (also known as starving a beach), placement loss for the beach itself under the rocks, and passive erosion which occurs because the back beach is fixed in space and can't go back landward, therefore, when sea level rises, the beach is submerged.[82]

How does sand accumulate on a beach?? It seems we are always talking about losses. Sand is transported over the seabed towards the beach when waves 'stumble' such that their crests become narrower than their troughs. This produces a swift forward flow followed by a slower backward flow. On the beach the top of the wave breaks and mingles with the foot of the wave, both dashing forward with a force driven by the energy from the height and speed of the collapsing wave. The resulting rush of water is fast and strong and moves sand effortlessly up the beach. The water then comes to a rest as the sand particles settle out. The water then begins flowing back down the beach, first slowly and then faster until it dislodges cohesive sand grains. But the forcing power is much less than that of the on-rushing waves. The particles that settled at the top of the forward rush stay because of the hysteresis (lagging behind) between erosion speed and settling speed. Thus, sand settles on the beach only when the tide recedes. Cuspate spits like you see in the harbor form when wind and waves come from along-shore, or at an angle to the beach. If wind comes onshore directly, cusps are less likely to form.

It is always helpful to collect several hundred grams of beach sand and bring it back for measuring through a series of sieves which sort the material into various grain sizes. The average amount of energy and wave and wind action that a beach endures is measured in its tiny grains of sand.

Large, poorly sorted, and angular (sharper edged) material indicates a beach that experiences high energy, large waves, and windy conditions. Fine sand that is well sorted indicates that the beach experiences relatively gentle waves that leave these finer particles behind. Max and Stefan also diligently sieved several samples of sand. That effort takes hours and hours and is exhausting. Yep, poor guys literally pounded sand. I helped them out for the last two-three hours. They found that the most common grain size of sand on the beaches in the intertidal areas in March and April was 500 micrometers (0.0195 inches) with 77% of that in the Cisco beach samples and 79% in the Codfish Park samples. The sand was finer and varied in size in the back beach stable dunes area where beach grass entrained finer particles. Next time you are out walking on a beach, reach down, grab some sand, and really look at it, take it home and look at it under a microscope or a hand lens, you'll see that these sand particles look like the boulders and rocks they came from.

Barrier Beaches: Nature's Fortresses
(June 27, 2013)

While writing about erosion on Nantucket over the years, I frequently felt that we did not show enough appreciation for the protection the harbor receives on the north and east from the long barrier beach known as Coatue and the

"Galls/Haulover." Everywhere you look on Nantucket you will find barrier beaches such as the ones lying between the ocean and Sesachacha, Hummock Pond, Miacomet, and Long Pond. Coatue, the "Galls", Great Point, and areas north of the Wauwinet gate house are part of a barrier beach system. For the twelve years I lived at the Nantucket Field Station, I'd stare out my dining room window at Folger's marsh and the barrier beach protecting it, in later years seeing storms overtop that barrier beach and remove copious amounts of sand built up over decades. During storms, barrier beaches act as a line of first defense, absorbing the wrath of the sea and protecting the land behind the beach. Marshes, dune, and beaches are components of a complex, restless system constantly interacting with wind and water to create the most dynamic of natural systems. Barrier beaches are especially fragile environments that are highly protected resources areas. Barrier beaches protect lagoons, estuaries, and salt marshes from the direct action of the sea.

There are three kinds of barrier beaches:

Bay barriers form between two adjacent headlands. The lagoons that are created by bay barriers may in time become freshwater ponds if the beach is not breached by the sea. This barrier is most common along the rocky coastlines of

New England and Canada. The spit of land between Sesachacha and the ocean is an example of this type of barrier beach. You can see lagoons becoming enclosed freshwater ponds. Nantucket's original harbor, Capaum Harbor, is now an enclosed pond bounded on the north by a barrier beach. Hummock Pond and Miacomet are protected by barrier beaches. These beaches do get breached in the larger storms and the Nantucket Field station barrier beach was severely breached during Hurricane Sandy in 2012 and the Nor'easter "Nemo" in 2013. Old pieces of a wrecked schooner and long buried student beach grass projects were unearthed in the erosion that took place at the field station with 3-4 feet of barrier beach height and 10 feet of width erased in the storms.

Barrier spits form where materials carried from a headland in tandem with a longshore current build a sandy "arm" into a bay or sound. Sandy Hook in New Jersey and Cape Henlopen in Delaware are examples of barrier spits. Local barrier spits include Smith's Point, Eel Point and Whale Island on Tuckernuck.

Barrier islands are the third type of barrier beach. They are long, low, usually narrow islands with inlets to the back-bay waters at either end. Barrier islands form a nearly continuous chain from New Jersey to Mexico. I lived for many years on the barrier island of Galveston, Texas.

In their natural state, free to move and react to the whims of the sea, barrier islands provide invaluable protection to the mainland. In the calm waters behind barrier islands, salt marshes, the most productive of all ecosystems, can develop and thrive. Barrier islands provide important habitat for many species of plants and animals. Left to the forces of nature, barrier islands are one of the most beautiful of all landscapes. Barrier island also tend to migrate toward the mainland by eroding on their high energy side (whatever direction faces the largest waves and ocean) and filling in on the low energy side. Barrier beaches such as Galveston and the long spits that used to lie to the south of Tuckernuck eventually join up with their "downwave" in this case northerly larger land mass.

From the Nantucket Conservation Foundations excellent website: "Nantucket's barrier beaches occur at the interface between the island and the surrounding sea. Coatue, Great Point, Coskata and the Haulover collectively comprise an ever-changing fragile strip of sand that shelters Nantucket Harbor from the open waters of Nantucket Sound and the Atlantic Ocean. These beaches have been deposited and reshaped over the last 6,000 years by ocean currents moving sand northward in a process called littoral drift, forming Nantucket's northernmost place – Great Point. An east-west current called longshore drift transported and deposited sand in a similar process to form the adjoining barrier beach known as Coatue. Smaller barrier beaches also occur at Smith's Point and Eel Point, on the southwest and northwest corners of the island."[83]

Over the years I received a lot of questions about beaches and ways to protect them, here are a few of those:

For years we have been putting Christmas trees on the shore to collect windblown sand. Is that a good idea?

Yes and no. In areas devastated by Hurricane Sandy in New Jersey, people are stockpiling Christmas trees and using them to augment dunes and beaches. The idea is not the world's worst, and the same tactic has been used on Nantucket for decades. Even the Department of Environmental Protection in NJ advocates for use of trees in small pilot projects. But coastal engineers and beach experts warn that these dunes are not as stable as naturally grown dunes made by beach grass plugs or even sand drift fencing. When a storm pulls out a Christmas tree beach, it tends to remove the whole structure and Christmas tree can prevent natural beach grasses from reestablishing. They are also fire hazards and can introduce invasive plants. The states of Delaware and Massachusetts no longer advocate for the use of Christmas trees for the above reasons.

The December 2008 Marine Extension Bulletin published by the Woods Hole Sea Grant & Cape Cod Cooperative Extension entitled "Coastal Dune Protection and Restoration: Using "Cape" American Beach Grass and Fencing"[84] also mention Christmas trees, but not in a positive holiday manner. They remind us that although it seems like a good recycling idea: "Discarded Christmas trees are not effective in maintaining a coastal dune under storm conditions. It has been shown that storm waves dislodge the buried trees, rapidly removing all accumulated wind-blown sand." But having said that, many communities in New Jersey and New York that survived Hurricane Sandy relatively intact did it with the help in some cases of long buried Christmas trees!

What type of plants can survive on a beach?

The Nantucket Conservation Foundation not only is the largest landowner on the island, protecting more than a third of Nantucket's land mass, they also provide an informative web site. Their description of common beach plants is excellent: "American beach grass is the most common plant found growing on the beaches of the Northeast. It forms an extensive network of underground stems. These rhizomes send up new shoots that hold windblown sand in place and promote the formation of new dunes. In the more protected areas behind the dunes, low shrubs that are able to grow under nutrient-poor conditions serve to further anchor the sand. Salt-spray rose, bayberry and beach plum are salt-tolerant shrubs that shed their leaves annually, adding nutrients to the sand as they decay. As the soil becomes more stable and enriched, other species such as eastern red cedar, black huckleberry and low-bush blueberry are able to colonize these areas, resulting in a diverse interdune plant community. Several rare and unusual plant species occur on Nantucket island's dunes, including the prickly pear cactus, oysterleaf and pink lady's slipper."[85]

Why is the sand different colors on the beach?

Beach sand is composed of several different minerals. Clear, glassy grains with uneven fractures are usually quartz, the most common mineral on the planet's surface. This mineral comes in several shades from clear to gray and black. Milky red, pink or white grains are usually feldspar. Glassy black or clear flakes are made up of mica. Most of the dark grains are varieties of hornblende. All beaches reflect the material of their parent rock (where they originated) and tend to become sorted in size based on the waves and energy of the beach. Higher energy beaches have coarser particles, up to the extremes of rocky beaches in which all smaller sand has been removed. Fine beach sand can only exist on beaches with gentle waves or extensive breakwaters.

What is the difference between a jetty and a groin?

Come on, get your mind out of the gutter! Groins are man-made structures designed to trap sand as it is moved down the beach by the longshore drift. A jetty is a large version of a groin built out into the ocean, on one or both sides of an inlet, to keep the inlet from filling with sand. Jetties, like groins, can create problems as well as provide benefits. Sand will build up on the "upcurrent" side of a jetty or groin because it is trapped by the structure, but sand will be scoured or removed on the "downcurrent" side because the longshore transport has been interrupted. There will always be beach erosion downdrift of the last groin. The point of jetties is to protect a major channel by minimizing shoaling and sand transport across the inlet between each jetty. But even inlets with jetties need periodic dredging. Groins are normally built to interrupt the transport of sand while jetties are built to protect a navigation channel or to keep a river or estuary from naturally meandering. They both are designed to stop natural processes of sand accretion and erosion.

Why do we have to stay off the dunes?

Planting American beachgrass (*Ammophila breviligulata*)[86] is the most effective way to stabilize existing dunes and build new dunes along our coastline. This vegetation is easy to plant and it spreads rapidly. It reduces wind velocity near the ground and traps windblown sand around the grass. As the sand deposits accumulate, the grass grows up through it maintaining a protective cover. This grass is not very tolerant of vehicle passage or people trampling it, so stay off those dunes! Beach grass is designed to snap off at the base, so only a few steps can result in beach grass dying, being blown out by the next storm and creating bald patches. Both the middle and northern Atlantic coast and much of the Pacific coast depend on this grass while Florida and the Gulf coast have a suite of other plants that build up their dunes.

The National Park Service's Guide to Cape Cod Beaches describes the role of beach grass in dune growth and protection eloquently and convincingly: "Beach grass is the creator and guardian of our sandy beaches and dunes. It is a perennial, tough, native grass that can withstand some flooding, salt spray, drought, strong winds and accumulating sand. It is the first plant to be seen growing on a forming sand dune. Bits of broken rhizome from the beach grass root system will start growing with little moisture. This is why beach grass grows back so quickly after a storm has torn it to pieces or buried it completely. When a healthy stand of grass develops, the stems break the force of winds and blowing sand. Grains of sand come to rest at the base of the stem and a good stand of grass can accumulate up to four feet of sand in a year's time. This natural system is more effective than artificial methods for dune building and rebuilding. When the root system of the grass is exposed, one can see clearly that the clumps of grass are connected to each other by underground horizontal stems, the rhizomes. These horizontal rhizomes have enlargements from which grow tough wiry roots that spread out into the sand. As the grass is buried and grows up, decaying parts of the grass create humus so other beach plants can take hold and grow. It is best not to walk on the grass with shoes, and not slide down or climb up a dune face. Wherever beach grass is destroyed, the dune begins to disintegrate and blow away."[87]

Thar She Blows
(Sept 7, 2011)

On August 29, 2011, I would have rather been writing about a giant white sperm whale. I was sitting in a hotel in

Hubbard, Ohio near the border of Pennsylvania desperately trying to get back to Nantucket to host an "Open House" event at the UMass Boston Nantucket Field Station. It is a common nightmare for island residents to be stuck either on or off island when you need to be on the other side. It is one of the many reasons islanders call the mainland "America."

I had perfectly good airline tickets burning a hole in my proverbial pocket, but I was driving because every east coast airport had been shut down in the wake of a massive storm racing up the Atlantic Coast. Hurricane Irene had barreled through and although it could have been a much stronger and more devastating hurricane, it has been an unprecedented storm in many ways. Mayor Michael Bloomberg evacuated a large portion of New York City for the first time in the days leading up to the storm. Many states, including those as far north as Vermont have taken unexpected hits due to the rain disgorged by huge multiple bands of thunderstorms from the spin art effect of the cyclonic storm. The storm brought torrential rains and powerful winds, stretching 300 miles from the center at one point and was frequently described as "the size of Europe" in various news media outlets.[88] Few storms in recent memory have affected such a large swath of the heavily populated coast. One consolation for people living in New England is that most storms run out of gas before they reach us, unlike the mid and southern Atlantic states. North Carolina's "Outer Banks" often take a beating from storms. Hurricane Irene washed out the lone road connecting the Outer Banks to the mainland, making it difficult for emergency crews to access the island with supplies and assistance. About 2,500 people on Hatteras Island were cut off from the mainland, and authorities sent a ferry Sunday August 28th full of supply trucks carrying food, water, and generators.[89]

Although hurricane season in New England officially begins on June 1, August and September are the prime months. Most of the 40 tropical systems that have hit over the past century have been in those months. Once a named hurricane causes significant damage; that name is retired. Hurricanes that produced massive rain or wind and storm surge damage to Massachusetts over the past 400 years include: Bob, Diane, Donna, Edna, Gloria, "Dog" (great name—probably after a labradoodle), Carol, the infamous "Long Island Express," "The September Gale (1869), the "Great September Gale" (1815), and a couple of others. The term gale was used for quite some time as the term hurricane was not widely known or accepted. The very first one recorded by colonists was known as the Great Colonial Hurricane of 1635 (Aug. 25, 1635). The storm's eye is believed to have passed between Boston and Plymouth causing at least 46 casualties.

That is not very many storms in such a long stretch of years. Fortunately for New Englanders, colder water and the sheering force of the storm grinding against the Atlantic coast weakens the majority of storms before they can do too much damage. In fact, the September 23, 1815 (the "Great September Gale" mentioned above) was the first major hurricane to impact New England in 180 years.[90]

The two most significant hurricanes to hit Nantucket over the past 30 years were Hurricane Bob and the "No-Name" or Halloween storm which assaulted the island in August and October of 1991 respectively. The No-Name Storm, as those who have enjoyed the book by Sebastian Junger or the movie "The Perfect Storm" know, formed as a low-pressure system built up over Canada, a high-pressure system extended from the Gulf of Mexico northeastward, and Hurricane Grace formed off the coast of Florida and moved north, all combining to create the perfect environment for a nor'easter to become a monster. Evidence for the "No Name" storm and "Hurricane Bob" can be seen in water marks two and three feet above the ground on the small office building at the UMass Boston Nantucket field station down by the water. Since then, more frequent storms have flooded that office to the point where it was permanently closed in 2014.

For Hurricane Irene, Nantucket Island was relatively prepared due to adequate early warnings, organized storm preparation, and the haul out of boats, preventing major damage to property. As is often the case in storms like this, the beach between Miacomet Pond and the ocean was breached, and south shore erosion and riptides were significant.

The National Hurricane Center (via the National Oceanic and Atmospheric Administration) is the best place to find comprehensive information about hurricanes. The terms "hurricane" and "typhoon" are regionally specific names for a strong "tropical cyclone." The term "tropical" refers to both the geographic origin of these systems, which form almost exclusively in tropical regions of the globe, and their formation in maritime tropical air masses. The term "cyclone" refers to such storms' cyclonic nature, with counterclockwise rotation in the Northern Hemisphere and clockwise rotation in the Southern Hemisphere. Hurricanes are known by different names in different places throughout the world. The storms we encounter on the east coast in the north Atlantic Ocean are called hurricanes. Storms that happen over the northwest Pacific Ocean are called typhoons. Hurricanes near Australia and in the Indian Ocean are called cyclones.[91]

The term cyclone derives from two different Greek words, "kuklos" meaning "circle" and "kuklōma" for "wheel" or "coiled snake" and was coined by British East India Company official Henry Piddington in 1848 to describe the devastating storm of December 1789 in Coringa, India.[92] It has been applied to tornados from 1856. The derivative of the word typhoon seems to have arisen from a combination of sources emerging in the late 16th century via Portuguese derived from Arabic ṭūfān with an assist perhaps from the Greek tuphōn for "whirlwind" and reinforced by Chinese word "tai fung" for "big wind".[93] The first evidence of the word hurricane occurred in the mid-16th century (a little earlier than typhoon) and came from the Spanish word huracán which is probably derived from Huracan/hurakán, or "god of storms" used by the Central American Tainos tribe.[94]

If you have spent any time watching the weather channel, you know that these bigger storms grow out of smaller "tropical depressions" which have maximum sustained surface winds of less than 34 knots or 39 mph. Once the tropical cyclone reaches winds of greater than that, they are called a "tropical storm" and are assigned a name. If winds reach 64 knots (equivalent to 74 mph), then they are upgraded to "hurricane" "typhoon" or "severe or Category 3 tropical cyclone" status. All these storms are characterized by a low-pressure center and numerous thunderstorms that produce strong winds and flooding rain. A tropical cyclone feeds on heat released when moist air rises, resulting in condensation of water vapor contained in the moist air. They are fueled by a different heat mechanism than other cyclonic windstorms such as nor'easters, European windstorms, and polar lows, leading to their classification as "warm core" storm systems."[95]

While tropical cyclones can produce extremely powerful winds and torrential rain, they are also able to produce high waves and damaging storm surge. They develop over large bodies of warm water and lose their strength if they move over land. This is the reason coastal regions can receive significant damage from a tropical cyclone, while inland regions are relatively safe from receiving strong winds. Heavy rains, however, can produce significant flooding inland, and storm surges can produce extensive coastal flooding up to 40 kilometers (25 mi) from the coastline. Although their effects on human populations can be devastating, tropical cyclones can also relieve drought conditions. They also carry heat and energy away from the tropics and transport it toward temperate latitudes, which makes them an important part of the global atmospheric circulation mechanism. As a result, tropical cyclones help to maintain equilibrium in the Earth's troposphere.

Scientists are looking at how climate change and our warmer sea surface temperatures might be affecting hurricanes by examining their frequency and strength over time. More energy in the system from warmer water (which is very well documented) means we must endure more dissipation of this energy in the form of more frequent and stronger storms. In fact, hurricanes do have a silver lining. In some areas like the Gulf coast, they can help to flush out contaminated harbors. Sediment layers in a coastal area will typically have an obvious "hurricane" layer where discontinuities in the deposition are observed.

One fun fact I found out researching cyclones and hurricanes is that Clement Wragge, the Australian forecaster who started the convention of naming tropical cyclones, occasionally named a particularly severe one after politicians with whom he was displeased.[96] That seems to be the perfect legacy to me.

The Trifecta of Moon Gazing Nights
(October 1, 2015)

On September 27th of 2015, I watched the total lunar eclipse on a lovely night with some cloud cover along with many people across the U.S. This one was special because the lunar eclipse occurred during a full moon, and a harvest moon, AND a "super moon": the trifecta of moon gazing nights. This meant that this full moon would be closest to the fall equinox, when the moon is closer to the earth ("super moon" or perigee moon) and hence looks bigger. The proximity of the moon in perigee on this September 27 event gave this event the grand sounding title, "super-moon eclipse." The last such eclipse happened in 1982, and the next won't occur until 2033. A lunar eclipse occurs when the earth blocks the sun from reflecting on the moon's surface which causes the moon to look blood red in color as red light is refracted around our dusty atmosphere and bounces off the moon's surface. If the Earth had no atmosphere, the moon would be completely dark during an eclipse.

A total lunar eclipse can occur only when the sun, Earth and moon are aligned exactly, or very closely so, with the Earth in the middle. Therefore, a lunar eclipse can occur only the night of a full moon. The term for this is being in "syzygy" and yes that was on my spelling word list the past few years. These lunar circumstances line up (if you will pardon a very bad pun) roughly every 20 years and have occurred five times since 1900. The most spectacular part of the eclipse will be the totality phase, when Earth's shadow completely covers the moon and turns it an eerie red. The moon will dip into the deepest and darkest part of Earth's shadow, or umbra, during the totality phase, which in this case lasted 72 minutes.[97] Many total lunar eclipses over the past few decades have been very short, sometimes only lasting five minutes, so this was an excellent chance to see one thoroughly. The type and length of an eclipse depends upon the Moon's location relative to its orbital nodes. Unlike a solar eclipse, which can be viewed only from a certain relatively small area of the world, a lunar eclipse may be viewed from anywhere on the night side of the Earth.

This weekend's blood moon will be the last in a series of four lunar eclipses, dubbed a tetrad, over the last two years. That pattern won't repeat for another 20 years or so.[98] Seeing this lunar eclipse reminded me of the equinoxes we experience each year.

In the language of science, an equinox is either of two points on the celestial sphere where the ecliptic and the celestial equator intersect. For the rest of us, it's one of two times a year when the Sun crosses the equator, and the day and night are of approximately equal length. For many people the autumnal equinox is a sign of fall and a harbinger of cool nights and approaching holidays. "Equinox" means literally, "equal night." As the angle of the earth's inclination toward the sun changes throughout the year, lengthening or shortening the days according to season and hemisphere, there are two times annually when day and night are of roughly equal duration: the spring and autumnal equinoxes. These celestial "tipping points" have been recognized for thousands of years and have given rise to a considerable body of seasonal folklore. At the autumnal equinox, the Sun appears to cross the celestial equator, from north to south; this marks the beginning of autumn in the Northern Hemisphere. The equinoxes are not fixed points on the celestial sphere but move westward along the ecliptic, passing through all the constellations of the zodiac in

25,800 years. This motion is called the precession of the equinoxes.[99]

The moon and lunar cycles are extremely important to biologists and oceanographers. During seasonally high tides more habitat is created for mosquitoes and crabs and fish as greater amounts of land is flooded. After viewing the eclipse, my first thought was that the color and translucence of the moon reminded me of an embryo or egg. That image in turn made me think of how dependent horseshoe crab eggs are on the tide which is influenced by the moon. Horseshoe crab spawning season varies according to latitude, but it generally peaks in May and June, with peak spawning occurring on evening high tides during the full and new moons (the higher-than-normal "spring" tides).

If you paid attention the week before a lunar eclipse occurs, you might have noticed the higher tides which became greater as the full moon approached. Tides are caused by the gravitational pull of the sun and the moon on the earth. The sun contributes about one third of the pull, and the moon two-thirds. The earth and the moon are attracted to each other in a gravitational sense and the moon pulls at the oceans.

Each day, there are two high tides and two low tides. When the moon is full or new, the gravitational pull of the moon and sun are combined. At these times, the high tides are higher, and the low tides are lower than other times of the month. The combined effects of this tide along with the higher tides of an equinox may produce minor flooding on the coast.

Sea turtles are also famously associated with the moon. For many years people thought that sea turtles needed the moonlight of a full moon to find their way to the ocean. This is in fact a myth. Sea turtles do go for the brightest light as they hatch from their nests and prefer to be born at night so they can more easily find the horizon and the starlight or moonlight reflecting off the sea. They will avoid the dark silhouettes of dunes or vegetation and flap cutely on their way to the sea. The colder temperatures of the sand at night is their cue that it is time to be born; under cover of darkness they have a better chance of surviving the short but dangerous journey to the surf. Their sea-finding behavior can take place during any phase and position of the moon, which indicates that hatchlings do not depend on lunar light to lead them seaward. One thing that scientists have observed is that the presence of lights from cars or oceanfront development can confuse the hatchlings and cause them to rush in the opposite direction.[100]

I have written about moths a couple of times in the past. Like clockwork, during this lunar eclipse one of my students repeated the adage that moths are attracted to light because they think it is the moon and once again a likely sounding story turn out to not be true. Moths do use the light of the moon to navigate but they do not head toward it. Dr. Mike Saunders, an entomologist at Penn State explains: "Moths often use the moon to orient themselves during night flight. In visual terms, the moon appears at "optical infinity," i.e., far enough away that the rays of light it reflects toward Earth are parallel as they enter a moth's (or a human's) eye. This constant makes an excellent navigational tool. Using the moon as a reference, moths can sustain linear flight in a given direction."[101]

We have a moth pollinated plant at the foot of my stairs, the evening primrose. The night blooming plant entertains my husband Len and the junior rangers at the Nantucket Field Station. The reason that evening primrose flowers open at night is to lure their primary pollinators, moths. The plants' flowers are "noctodorous,"[102] they release a subtle fragrance at night to lure their moth pollinators close.

For once, this lunar eclipse lived up to the hype and was not blocked by clouds or fog. Relatively clear skies with just a few suspense-buildings clouds allowed us to observe the cloaking and uncloaking of a maroon-orange orb of light 221,753 miles (356,877 km) away.

Restoring Our Marshes
(July 31-Aug. 6, 2008)

Decades of human manipulation have transformed our salt marshes and wetlands. Formerly considered as worthless water-sodden land that bred diseases, mosquitoes, and monsters,[103] many marshes and swamps were filled in or drained. Entire neighborhoods were built on filled tidelands along the east coast, with Boston's Back Bay as a prime example.

In the mid-1930s, during the Depression, the government put people to work in programs designed to increase employment through agencies such as the short-lived Civil Works Administration and the much more successful WPA (Works Progress Administration, later renamed the Works Projects Administration). One of the jobs created was ditch-digging in salt marshes. More than 90% of all east coast salt marshes were ditched into grids, with close to 3000 miles of ditches dug in Massachusetts alone.[104] This ditching was originally designed to eliminate mosquito-breeding habitat on the marsh by draining the water off the marsh quickly after high tides.

In the short term, this worked while the ditches were clear of infill. One of the unforeseen side effects, though, was that once the marshes were drained, they were no longer used by wildlife. Another problem was that the ditching was not always maintained and in some situations the ditches that were designed to prevent mosquito breeding, became prime mosquito breeding habitat as the water in the ditches stagnated. Ditching was also done to facilitate salt marsh haying and in some areas fish weirs were established by the native Wampanoags that diverted natural tidal inlets.

For the past decade or two, federal, state, local, and private groups have been using techniques such as Open Marsh Water Management (OMWM) to restore New England marshes to their former glory. OMWM (how would bureaucracies manage without acronyms and initialisms?) involves filling in some ditches, reconnecting other ditches to deeper pools, and recreating salt pannes in a marsh system to allow normal tidal flow and mosquito-larvae-eating fish such as *Fundulus* (killifish) to reach all areas of a marsh where mosquitoes might breed. Basically, OMWM is undoing the work that 11,000 men did in the peak grid-ditching year of 1934. OMWM also allows native marsh vegetation such as *Spartina alterniflora* (smooth cordgrass or saltmarsh cordgrass) and *Spartina patens* (salt meadow cordgrass) to fill in naturally.[105] All these efforts are an attempt to restore normal tidal flow. The tide is the dominating characteristic of a salt marsh. The salinity (salt content) of the incoming tide defines the plants and animals that can survive in various sections of a marsh. The vertical range of the tide determines flooding depths and the height of the vegetation, and the tidal cycle controls how often and how long vegetation is submerged. Two areas are delineated by the tide: the low marsh and the high marsh. The low marsh generally floods and drains twice daily with the rise and fall of the tide; the high marsh, which is at a slightly higher elevation, floods less frequently.

The salt marsh is one of the most productive ecosystems in nature, with the total biomass created daily rivaling a typical rainforest. In addition to the solar energy that drives the photosynthetic process of higher rooted plants and the algae growing on the surface muds, tidal energy repeatedly spreads nutrient-enriched waters over the marsh surface. Some of this enormous supply of live plant material may be consumed by marsh animals, but most of the vegetation dies and is decomposed by microorganisms to form detritus. Dissolved organic materials are released, providing an essential energy source for bacteria that mediate wetland biogeochemical cycles (carbon, nitrogen, and sulfur cycles). A healthy marsh typically increases in height gradually (by millimeters) as this sediment accumulates; this provides a fragile balance between subsiding coasts and increasing sea levels. The salt marsh serves as a sediment sink, a nursery habitat for fishes and crustaceans, a feeding and nesting site for waterfowl and shorebirds, a habitat for numerous unique plants and

101

animals, a nutrient source, a reservoir for storm water, an erosion control mechanism, and a site for aesthetic pleasure. Appreciation of the importance of salt marshes has led to federal and state legislation aimed at their protection.

While we are on this topic, it has always bothered me when someone calls a marsh a "swamp." Frankly, I am never a fan of draining swamps, they are there for a reason and all that does is denigrate a swamp. The basic distinction between swamps and marshes depends on whether the wetland contains trees. Swamps are forested wetlands, containing trees and large shrubs. Marshes, on the other hand, are primarily filled with grasses and various soft-stemmed plants. Swamps also often have more open water and tend to be deeper than marshes. I enjoy both swamps and marshes, but I have found that when someone uses the term "swamp," it is not meant as an endearment.

One of the many avoidable tragedies of Hurricane Katrina's assault on Mississippi and Louisiana was in areas where salt marshes were drained to make canals or filled in and subsequently built up. America's Wetland, a Baton Rouge organization, estimates that more than 1,900 square miles of the Louisiana wetlands (approximately 25% or 1.2 million acres) have disappeared between 1932 and 2000 due to development and the construction of levees and canals.[106] These areas could have helped to absorb the storm surge that came as far as 50 miles inland in some areas by providing a frictional force that slows down the flooding water. Scientists have estimated that for approximately every square mile (640 acres) of salt marsh filled in, another foot of storm surge occurs.[107] Now, as sea level rise is overcoming sinking or dredged wetlands, we are starting to value, both scientifically and economically, the worth of this land.

On Nantucket, conservation groups have been proactively working to maintain and enhance the natural functioning of salt marshes for many years. The Nantucket Conservation Foundation (NCF) in collaboration with the Massachusetts Office of Coastal Zone Management (CZM) Wetlands Restoration Program (MWRP) and the University of Massachusetts at Boston Nantucket Field Station conducted research on the Medouie Creek marsh system (2004-present) to determine how to design a restoration plan, implement it, and then evaluate its effectiveness to counteract or slow the establishment and spread of the invasive common reed (*Phragmites australis*) within the marsh. The spread of Phragmites in our nation's wetlands has pushed out native plants and the animals that depend on them through the establishment of a monoculture which supplants the natural biodiversity of the wetlands. The next time you drive along coastal areas in New Jersey, Rhode Island, Connecticut, or Massachusetts, look along the roadsides for this 10 to 20-foot-tall reed to understand the magnitude of the problem.

Some severely impacted areas can never be reclaimed, but it is worth the effort for areas such as Medouie Creek marsh. In 2003, the MWRP designated Medouie Creek as a high priority restoration site due to the value of its habitat and the potential for restoration. Medouie's tidal exchange with the harbor had been severely restricted by the presence of dike roads and the partial blockage of the only remaining connecting channel. From 2003-2008, the NCF, with some help from the Nantucket Field Station, did an extensive amount of research work including installing pressure transducers through the marsh system to measure tidal height, measuring soil salinities at various depths, obtaining exact elevation measurements, and doing vegetative profiles to document the "before restoration" conditions. NCF also hired hydrologists who study water movement, properties, and effects, and worked with the CZM to determine the best plan of action based on their data and all alternatives.

The NCF project restored substantial tidal exchange by installing a meter square box culvert at a low point in the dike road and by opening up channels that originally led to the southern end of the marsh before the dike road was built. Their project allowed salt water to reach the back portions of the marsh and improve water quality while holding the insidious *P. australis* at bay. It also reduced the impoundment of freshwater which could not easily drain after high rainfall events. The combined influence of increased freshwater drainage and saltwater inundation are predicted to improve habitat conditions for native salt marsh flora and decrease habitat suitability for non-native *P. australis*. Thankfully, on Nantucket, we know our marshes are more than just useless land.

Section 2: It's a Hard-Shelled Life

What the heck is monkey dunk?

(originally published as "Sponges, Climate Change, and Bay Scallops" on August 8, 2013)

Boring Sponge (photo by Elizabeth Boyle, PhD)

On a typical summer day in 2013, we pulled up a common resident of our harbor while doing a dredge for the UMass Boston Marine Ecology class off Pocomo Point. Scattered amongst the spider crabs, bay scallops, loose eelgrass and assorted algae was the omnipresent boring sponge. The large gloppy yellowish blobs are called "monkey dunk" by scallopers who dredge them up in the winter. What we hauled up has also transformed many of the shells you will find on the beach. Next time you are walking at the beach, look for shells with dozens of small, almost perfectly round holes in them. Those were made by a boring sponge named for its penetrating skills (not monotonous conversation) that uses acid to dissolve the calcium-based shell of its host. The only function of the shell is a place to live (a substrate) for the sponge. The shellfish living in the drilled shell is not consumed, but usually dies since its protective covering has been damaged. Thank goodness bivalves don't have higher cognitive functions because I think it would be scary to know something is slowly eating its way through your home.

Sponges are the oldest living multicellular organisms, with a fossil record dating back to about 600 million years.[108] They are often considered to be the most primitive of animal groups, largely due to their simple body structure that lacks actual tissues or organs, yet they represent a highly specialized and successful group of benthic suspension-feeders in marine and fresh waters (not to mention their own cartoon shows). So adaptable are sponges that a small group in a food desert in the deepest ocean has successfully switched to a carnivorous lifestyle from their normal filter feeding modus operandi.[109] Their persistence throughout geological time, wide global distribution, and occurrence in diverse habitat types are more evidence of their adaptability. I am not saying that most life will be some type of sponge in 20,000 years, but it is possible,

The sponge phylum, Porifera which means "the pore bearers," is divided into three distinct classes. The Calcarea have spicules (tiny spike-like structures) of calcium carbonate in the calcite form and are commonly known as "calcareous" sponges which makes sense. The Demospongiae, which is the most diverse class of sponges, have skeletons of two components, a mineral skeleton of silica (glass) spicules, and an organic skeleton made of diverse forms of spongin (collagen) fiber. Some sponges lack spicules, some lack fiber, and some combine the two. Some even incorporate sand into their skeleton. Some demosponges, called the "sclerosponges," form a stony base of calcium carbonate in the aragonite form, with a veneer of living "tissue" on the surface. These sponges also produce silica spicules. Lastly are the Hexactinellida, commonly known as the glass sponges, which have siliceous spicules with a hexactine (six-ray) design.[110]

According to the real UMass Boston Nantucket Field Station marine biologist (unlike me), Dr. Beth Boyle, who teaches the advanced level of Marine and Coastal Ecological Research at the Nantucket Field Station each summer and is an expert on invertebrate species, the species name of the huge pile of yellow sponges we found (several pounds) is most likely *Cliona celata* which is a demosponge. She told me that to positively identify a sponge you would normally do a spicule analysis under a microscope to confirm that you have the right species. *Cliona celata* is a cosmopolitan species which means it is well traveled for a sessile marine species and it can be found anywhere in the world with an appropriate habitat (reminds me of a "Sex in the City" character). It starts out boring into shells or even limestone and can eventually overgrow its host and reach the state we saw it on the boat where the host was no longer visible. Many of the sponges we broke open only had a few sand particles inside them. Sponges feed by pumping water through their pores. Most sponges work somewhat like chimneys: they take in water at the bottom and eject it from the osculum ("little mouth") at the top. Since ambient currents are faster at the top, the suction effect that they produce does some of the work for free. Sponges can control the water flow by various combinations of wholly or partially closing the osculum and ostia (the intake pores) and varying the beat of the flagella and may shut it down if there is a lot of sand or silt in the water.

Demospongiae can form thin encrustations, lumps, finger-like growths, or urn shapes. Most are marine dwellers, but a few live in freshwater environments. Pigment granules in amoebocytes often make members of this class brightly colored, including bright yellow, orange, red, purple, or green. There is great variety in body size from just a few millimeters to 2 meters (3.3 ft) across. These are asymmetrical sponges.[111]

Our local "monkey dunk" can bore through solid rock (provided it is limestone), like the others in the genus

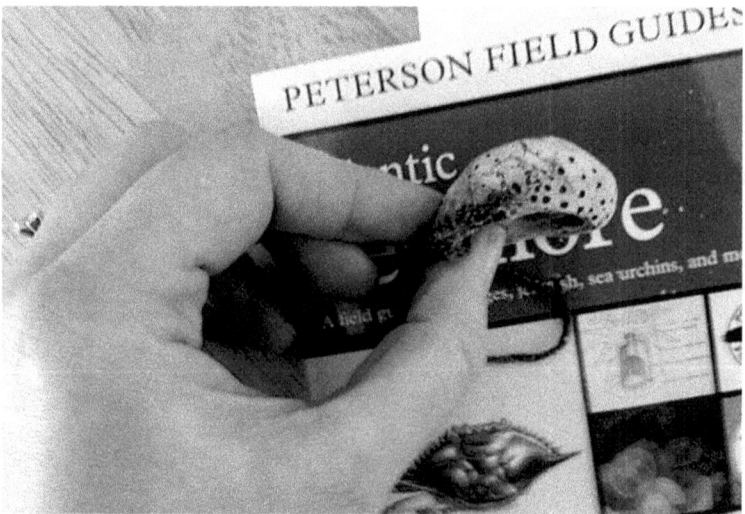

Photo by the author.

Cliona. *C. celata* can be up to 100 cm across and 50 cm tall but are usually smaller. They have a coloration of different shades of yellow, but in some cases, red coloration around the openings which gives them the name "red boring sponges." They can adapt to many environments by changing the way their cells are distributed across the substrate. These forms can be ridged, burrowing, massive, massive/chimneys, encrusting, or encrusting/chimneys. It is typically easy to recognize *C. celata* because of its color and sieve-like openings. This sponge can make round holes up to 5 cm in diameter in limestone or the shells of mollusks, especially oysters. They seem to really like large quahogs here and you will often find densely bored quahog shells on our beaches. The whole sponge shows a noticeable decrease in size when removed from the water. The consistency of this sponge is firm and inflexible with a tough outer layer that often feels almost fake or plastic-ish. They feel like somewhat like old buoy material to me.

In the world porifera database,[112] we find out how *C. celata* got its name. The sponge was first described by R.E. Grant in 1826 in an article called "Notice of a New Zoophyte (Cliona celata Gr.) from the Firth of Forth".[113] Love that location name, the "firth of Forth", sounds like a Monty Python gag. "Celatus" is Latin for "concealed or hidden," referring to the boring habit, and "Cliona" is from the New Latin for Clio, the Muse. She is known as the Muse of History and is often shown carrying parchments or scrolls and is sometimes known as the Proclaimer. From a pile of yellow gunk on your deck to the muse of poets and historians, now you see why I like etymology.

The boring form is very common in oyster and mussel beds, where they cause some damage to shellfish farming. Some of the boring sponge's favorite victims include: *Crassostrea virginica* (eastern oysters), *Crepidula fornicata* (Slipper limpets) and *Mercenaria mercenaria* (hard clam or northern quahog). It may be confused with the similarly colored *C. lobata*, but that species has much smaller and more numerous papillae, and it has tiny spicules called microscleres (*C. celata* do not). Sometimes *C. celata* is also confused for or listed in earlier works as *Cliona sulphurea* (Disor), which was found in an early 20th century Maria Mitchell Publication.

There are some other species of *Cliona* found near coral reefs that are indicators of excessive organic pollution, but this species can be abundant in perfectly pristine waters. Some of the reasons that "monkey dunk," aka this boring sponge is common around Nantucket include, plenty of shells for them to colonize, decent water circulation in the harbor to extract food from (suspension feeders), and a lack of predators. Not many things eat sponges, their main predators are nudibranchs, small crabs, and small crustaceans, but they can regenerate cells lost to small amounts of predation (minor "flesh wounds"). This sponge can reproduce asexually and sexually. They can simply

111

separate by mitosis, as single cells do, or they can release sperm into the water in hopes of them finding a female's eggs.

The effects of climate change are being studied for a multitude of invertebrates including the many species of bivalves we eat. The effects from these environmental changes will vary among taxonomic groups, with calcifying organisms like corals and mollusks considered to be most at risk of adverse impacts like shell loss, body weight reductions or even inability to stay alive.[114] Scientists Alan R. Duckworth and Bradley J. Peterson mimicked conditions that are likely to occur in coastal waters in the next fifty years as the ocean's waters warm up and become more acidic from increased levels of carbon dioxide in the atmosphere. The warmer temperatures did not do much to the boring sponges, but the lower pH (7.8 versus 8.1) water made a huge difference. The sponge's rate of bioerosion (how it drills into the shell) doubled in the more acidic and warmer conditions resulting in scallop shells that were significantly weaker and twice as thin as normal which exposes them to greater predation.[115] Sometimes it is hard to see on a local level what climate change can do, but this is certainly a sobering thought. For many years I measured pH frequently in the harbor and it was relatively stable at 7.8 pH which matches the experimental pH used in the paper above so our harbor sponges may already have adapted to more acidic and warmer water. For comparison, open ocean water worldwide used to have an average pH of 8.2 and that has dropped to 8.1 since the beginning of the Industrial Revolution. The oceans are doing a substantial amount of carbon dioxide absorption sucking down 30% of our emissions.[116] One possible saving grace is that the scientists also found that the sponge was taking longer to attach to shells at the lower pH and slightly warmer conditions. Sponges attach to a substrate using a collagenous matrix made of spongin and fibronectin (both good scrabble words) that is secreted by cells on the base of the sponge called basopinacocytes. The authors theorized that the low pH might interfere with this secretion process.[117] We can look for any port in a storm, as they say. Next time you are at the beach, look closely along the shore for the remnants of the boring sponge's attack on its aquatic neighbors.

Hermaphrodites in the Harbor
(May 20, 2010)

During a trip to one of the island's beaches, you'll quickly find one of the most common snails on the island, the slipper shell; also known as the slipper limpet or mermaid's slipper. The common slipper shell, *Crepidula fornicata*, has a host of other names (and for all we know the passports to match) including common Atlantic slipper snail, boat shell, quarterdeck shell, and is known Fin Britain as the "common slipper limpet" although it is not technically a limpet.[118] This is a species of medium-sized sea snail, a marine gastropod (Phylum Mollusca) in the family Calyptraeidae, which includes the slipper snails, the Chinese hat shells (*Calyptraea* species), and the cup-and-saucer snails (*Crucibulum* species). The Atlantic slipper shell is arched with brown or pink markings and an apex that is bent downward to one side at the back like an angled corkscrew. They live in relatively shallow water to depths of about 20 feet. The shell is oval, up to 5 cm in length, with a much-reduced spire and no operculum or "door" that closes the shell and protects the creature inside. The large opening (see picture below) has a shelf, or septum, extending half its length.[119] On Nantucket one can also find, in much smaller numbers, another species of slipper shell called the *Crepidula plana* or Eastern white slipper shell, which is flatter than the *C. Fornicata* and has a shorter septum. Unlike a bivalve (two-part shell) such as the clam or oyster, the slipper shells have a one-part shell with the creature's foot on the underside not unlike sea and land snails. Slippers shells' whimsical shape and resemblance to household slippers (for very tiny people) make them a favorite beach combing find for children.

The *Crepidula* species and other members of the family *Calyptraeidae* are sequential hermaphrodites[120], which mean they can alter their sexual identity. The largest and oldest animals at the base of a pile are female; the younger and smaller animals at the top are male. If the females in the stack die, the largest of the males will become a female. Try describing this phenomenon in a responsible way to a group of youngsters, good times, lots of giggling. To make it even more interesting, *C. Fornicata* are protandrous hermaphrodites, meaning they always start out male, they would be called protogynous hermaphrodites if they were female first. When the larva first settles on a substrate, it is male and it bides its time until something tells it to switch to a female and then it sends out signals to attract others to land and attach on top, otherwise it would be lonely, and unproductive.

This sexual division of labor makes it extremely easy to remember the Latin name of this creature if you catch my drift, wink-wink, say no more. Eggs are laid (70-100 at a time) in thin-walled capsules which are attached to the substrate (the base on which the creature lives). Although breeding can occur between February and October, peak activity occurs in May and June when 80-90% of females spawn. Most females spawn twice in a year, usually after neap tides (a tide just after the first or third quarters of the moon when there is least difference between high and low water). Females can lay around 5000-10,000 eggs at a time.[121] That is a feat that is hard to believe out of such a tiny creature. Many shellfish are extremely dependent on releasing large numbers of offspring to successfully reproduce.

Incubation of the eggs takes 2-4 weeks, followed by a planktotrophic larval phase lasting 4-5 weeks. A long planktonic phase of 4-5 weeks provides plenty of time for the larval snails to find a place to call home. Recruitment (where the spat will settle) is determined by the local hydrographic regime, in layman's terms, how the bottom of the bay or harbor is shaped and where the currents flow. For example, in sheltered bays, the larvae may be entrapped in circular currents and small-scale eddies (or whirlpools) which can result in the concentration of larvae. The ability of *Crepidula fornicata* to disperse widely and colonize new areas is demonstrated by its spread through Europe following introduction from North America at the end of the 19th century. The spat can settle either in isolation or on top of an established chain of *C. fornicata*. *C. fornicata* needs to be part of a chain in order to breed and therefore would be expected to settle preferentially where high densities already exist. Males reach sexual maturity two months after settlement. If a male develops directly into a female, sexual maturity may be reached in 10 months.

Although slipper shells are members of the snail family, they are very different from most snails. They have a relatively flattened shape versus the spiral shape the majority of snails' exhibit. These snails split off from the rounder snails you might be familiar with between 20 and 100 million years ago and have adapted to high wave environments. Instead of traveling around on a strong muscular "foot," slipper shells remain semi-permanently attached and filter feed in place by lifting their body from the rock, other slipper shell, or horseshoe crab they have latched onto. This ability to stay in place on a substrate allows slipper shells to avoid drying out, otherwise known as desiccation. There is some evidence that the juveniles (always restless, even as gastropods) will use their radula (like a tongue) to scrape algae from the substrate. You'll also find slipper shells attached to other shells along the beach, such as whelks, or to each other in stacks. Brant

Point is literally covered with empty slipper shells and you will also find them stacked together on the bottom of clumps of *Codium fragile* otherwise known as "oyster-thief" or "dead man's fingers."[122] In storms, the more concave or cup like slipper shells are dislodged by wave action while the flatter shells remain firmly attached.

The genus *Crepidula* is probably the best studied group of calyptraeids. A variety of species are commonly used in developmental, ecological, and behavioral research. They have been the major focus of research on protandrous (male to female) sex-change in marine invertebrates and it has very recently been discovered that touch with other snails is required for this change to occur, not water borne chemical stimuli.[123] *C. fornicata* is a well-studied example of an invasive, exotic species in marine habitats. Slipper shells are a serious invasive species in Europe. They were introduced via oyster farming and are sometimes called American slipper shells in the scientific literature, just to make it clear what country is thought to be responsible for the invasion. The species is considered an invasive species in Denmark, France, Italy, the Netherlands, Spain, and the UK, and has also spread to Norway and Sweden. It is known to damage oyster fisheries. It has also been introduced to the Pacific Northwest and Japan.

Because slipper shells attach themselves to a hard object in the water and spend their lives in that one place filtering macroalgae from the water using large gills, they are often studied to evaluate their tolerance to various pollutants. They are extremely intolerant to petroleum products and most surfactants, but able to live in areas inflicted with harmful algae blooms like the brown tide which decimated Peconic Bay scallops.[124]

Scientists have also investigated whether *C. fornicata* can filter out excess nutrients and sequester (hold onto) various metals. They are somewhat tolerant of metal contamination, living in bays with high levels of pollution but at reduced sizes. Their reproduction cycle is short, so they can recover quickly after bouts of mortality.

One of my visiting Boston area high school students was determined to find out whether each animal and plant he encountered on our nature walks could be safely eaten. He hit the jackpot with the slipper shell which is known locally as sweetmeat and often eaten. I am not sure his classmates were as enchanted when he brought a pocketful home to be boiled and consumed, but slipper shells can be safely eaten boiled or raw immediately after harvesting. I will admit I have not tried them. With a little bit of looking, I found some recipes online.[125] Bon appétit!

One final fascinating tidbit about these creatures was unearthed when researchers opportunistically used the presence of slipper shells on horseshoe crabs to determine the age of the crabs, which can be difficult to do. From this research and additional tagging studies, they estimated that horseshoe crabs live 9 to 12 years before maturity and 5 to 7 years as adults, for a total lifespan of 14 to 19 years and slipper shells live 8-11 years. [126] Not bad for a snail!

Feeling Crabby?
(Sept 7, 2010)

At the Nantucket Field Station, I kept two aquariums full of creatures captured in Folger's Marsh and in the harbor including both spider and blue crabs. These crustaceans were the source of much amusement, some horror, and gobs of educational mayhem. Each fall they were released back into the marsh and harbor to tell their stories of alien abduction.

True crabs are decapods ("ten footed") crustaceans of the infraorder Brachyura[127]. They got their name from having a very short projecting "tail" which is in fact their reduced abdomen which is hidden under the thorax. The name comes from the Greek βραχὺς, romanized to "brachys" for "short" and "oὐρά/oura" for "tail",[128] Other animals, such as hermit crabs, king crabs, porcelain crabs, horseshoe crabs and crab lice, are not true crabs. There are many other crabs that call Nantucket's waters their home; both native species and interlopers paddle, skulk, scamper, skitter, or crab-walk around. Common species include the lady crab, black-fingered mud crabs, spider crabs, blue crabs, and fiddler crabs, less common are the rock crab and the shame-faced crab. Invasive species include the green crab and the Asian/Japanese shore crab. Out of all the crabs you'll find, I think the blue crabs and spider crabs are some of the most intriguing; they seem to represent the yin and yang of the crab kingdom.

Both the constellation Cancer and the astrological sign Cancer are named after crabs. John Bevis first observed the Crab Nebula and its resemblance to the animal in 1731.[129] The word "cancer" evolved from the Greek word "carcinos" used by Hippocrates (460-370 BC) to become the Latin word "cancer" when used by the Roman doctor Celsus (28 BC - 50 AC). The term was applied to tumors because these physicians felt tumors looked a bit like a crab.[130] I don't see the resemblance, but I am not a medical doctor.

Many crabs exhibit sexual dimorphism which means that creatures show some difference physically between the sexes. A common example is a larger claw for males to attract mates and deter rival male crabs. Crabs exhibit physical differences in their "apron" or abdominal area because the females carry the eggs under themselves where they can protect them with both claws. To determine the sex of a crab without buying one a drink, you need to flip them over and look at their bottom shell or "pleon." All crabs have a somewhat triangular abdomen, but the females have rounder, larger, broader abdomens while the males' abdomens are narrow and pointed to form a "lighthouse" or "T" or a "Washington Monument" shape. The patriotic mnemonic helps significantly; if the abdomen looks like the Washington Monument, the crab is male; if it looks like the U.S. Capitol, it is female. You can draw all the conclusions you want from that. Moving along... Another helpful clue to the gender of a blue crab is that female blue crabs have red tips (i.e., they "paint their fingernails") on their claws while the males have blue tips, which is good to know as opposed to trying to flip over a crab to see its apron when it is not in the mood. The front "claws" or pincers are called the "chelipeds" from the ancient Greek word "khēlé" which evolved into the new Latin word "chela" (always competing for etymological supremacy) for "claw." [131]

Crabs are omnivores and will eat anything available from algae to detritus to other crabs to worms, oysters, clams, and other bottom dwelling invertebrates. They do best on a mixed and varied diet. They are preyed upon by a variety of species from fish to other crabs to whelks.

The scientific/Latin name for the Atlantic blue crab is *Callinectes sapidus*, from the Greek words "calli" for "beautiful," "nectes" for "swimmer," and the Latin word "sapidus" which means "tasty" or "savory." Blue crabs are found in the waters of the western Atlantic Ocean, the Pacific Coast of Central America, and the Gulf of

Mexico. They have been introduced to parts of the Eastern Atlantic, in the northern and eastern Mediterranean, and in Japan. Blue crabs are extremely important to the scallop fishery on Nantucket because they tend to predate on green crabs. Green crabs eat larval species of scallops. They also feed on eelgrass, which is an important habitat for the bay scallop. There is evidence that blue crabs in eastern North America can control populations of the invasive green crab, *Carcinus maenas*; numbers of the two species are negatively correlated, and *C. maenas* is not found in the Chesapeake Bay, where blue crabs are most frequent.[132]

All crabs have exoskeletons, which mean they must molt to get larger. Female blue crabs mate only once in their lives, when they become sexually mature immediately following their pubertal molt (immediately following this molt, the female is known as a "sook" and her abdomen looks more rounded versus a triangle). When approaching this pubertal molt, females release a pheromone in their urine which attracts males. Male crabs compete for females and will carry and protect them, called "cradle carrying," until molting occurs. Following this molt, when the female's shell is soft, the pair will mate. During mating, the female captures and stores the male's sperm in sac-like receptacles so that she can fertilize her eggs later. Once the female's shell has hardened, the male will release her, and she will migrate to higher salinity waters to spawn. It looks sweet, but essentially is a hostage taking situation.

The frequency of spawning depends on latitude, with two spawning periods (spring and summer) for crabs in the Chesapeake Bay which live typically for three years. In the wintertime, crabs are dormant and stay buried in the warmest, siltiest areas they can find. Mating occurs primarily in relatively low-salinity (fresher) waters in the upper areas of estuaries and lower portions of rivers. Mating takes place in areas where female crabs normally go to molt, typically shallow areas with vegetated banks or beds of submergent vegetation, [133] my guess is it is more private that way. Mating for blue crabs is a very involved and drawn-out process and fascinating. One year, the Nantucket Field Station hosted a mating pair of blue crabs and it was an amazing sight. We also got to see some cannibalistic behavior which is common for blue crabs. Cannibalized blue crabs make up as much as 13% of a crab's diet. Blue crabs in poor health, missing legs, heavily fouled with other organisms, and those during or immediately following a molt and unable to defend themselves are more likely to be cannibalized.[134] They don't even wait to become zombies to eat each other's brains.

Crabs hatch into a zoea stage which bears no resemblance at all to the adult. Crab zoea will eat a variety of plankton (both plant and animal). There are usually seven zoeal

124

stages and one postlarval, or megalopal, stage although if the temperature and salinity change, there may be fewer stages. The zoea can only get larger by molting. The next stage is the megalopa stage where the crab starts to kind of look crab-like, although their abdomen is extended which makes them look like a little shrimp or copepod. The megalops stage lasts 6 to 20 days, after which the megalops molts into the "first crab" stage, with proportions and appearance more like those of an adult.[135]

Although they can live in a variety of salty and semi-fresh or brackish waters, blue crabs tend to be a bit more sensitive to temperature and pH (acidity). When air temperatures drop below 50°F (10°C), adult crabs leave their shallow close to shore water and seek deeper areas where they bury themselves and remain in a state of torpor throughout the winter.[136] Blue crab growth is regulated by water temperature. Growth occurs when water temperatures are above 59°F (15°C). but if the water gets too hot, above 91°F (33°C) they will die. One of the few positive responses to warming water temperatures is that modelling shows that blue crabs will be more likely to survive the winter in the Chesapeake Bay area and they may also grow faster.[137]

Blue crabs are also affected by pH and salinity to some extent depending on their life stage, as they get older, they are less susceptible to water chemistry changes. Tomasetti et al (2018) found significant reductions in zoeal survival after long-term (fourteen day) exposure to low pH conditions ranging from 7.2–7.32.[138] As is the case with many marine invertebrates, climate change brings both positive (warmer water) and negative (lower pH) factors to the blue crab life cycle equation.

The blue crab forms the basis for a lucrative fishery. A large component of the fishery in Maryland and many Southern states is the soft-shelled crab harvested immediately after molting. The harvest of blue crabs has stabilized after strict fishing limits and protections for female crabs were put into place in 2008. In 2021, juvenile and male crab surveys showed a depleted population, but female populations were stable.[139]

With warming ocean waters, blue crabs are also moving north. Research David Johnson first documented the northward movement of blue crabs into the Gulf of Maine in 204 and 2015 and attributed it to warming waters.[140] In August 2021, a UMass Boston biologist and long-time friend of the show, Dr. Jarrett Byrnes also reported seeing a live blue crab in Maine, north of its traditional range. As climate change causes warmer sea temperatures, these may be one of the animals that will benefit with a longer growing season and larger range.

Spider crabs, although they look much fiercer than blue crabs, are relatively mellow (for crabs) and much less aggressive to their fellow tank mates. There are two species found in the waters surrounding Nantucket and up and down the North Atlantic coast, the portly spider crab (*Libinia emarginata*, Leach 1815) and the long-nosed spider crab (*Libinia dubia* H. Milne Edwards 1834). I am sure the portly spider crab, which is also called the common spider crab and nine-spined spider crab, would call itself "big-boned."

Both crab species have long thin walking legs springing out from its rounded body which explains the name "spider crab." These legs culminate in curved points, allowing the crabs to cling to various surfaces from rocks to docks. *L. emarginata* is triangular in outline, with a carapace about 4 inches (100 mm) long and a leg span of 12 inches (300 mm). These are big, very awkward, and strangely endearing creatures. They look like a softball balanced on spindly legs that came out of Jim Henson's (creator of the Muppets) fertile mind. The typical crab is khaki/mud-colored which is excellent camouflage on the ocean sediments, and their carapace is covered in spines and tubercles (warty knobby outgrowths) and is often clothed in debris and small invertebrates (for even better camouflage).[141]

For *L. emarginata*, mating takes place, and eggs are produced from June to September. They mate in large groups and the males often hold gravid females behind them and protect them aggressively.[142] The eggs are initially a bright orange-red, but turn brown during development, which takes around 25 days. The eggs then hatch as zoea larvae, and the female can produce another brood of eggs within 12 hours, unlike many other crabs, where females only mate directly after molting and then have their next brood many months later.[143] It is a bit tough to tell these two apart, but the portly/common spider crab is rounder, and the long-nosed spider crab is teardrop shaped and does have a longish "nose" or rostrum area. Less is known about spider crabs because they are not a commercial species.

The portly spider crab is also one of the few crabs to walk forward as opposed to the sideways motion which is easier for most crabs. These crabs are scavengers and hunt by smell. They also can eat much larger prey (sessile prey) like the Forbes Sea Star (*Asterias forbesi*).[144]

The blue crab has modified back legs into swimmerets or paddle legs which help them swim where the spider crab has pointed appendages. Remembering their order of Decapoda, they both have 10 "legs" which are modified to serve a variety of needs. *L. emarginata* are slow moving and have sense organs for taste on the tips of their legs that can detect food in the mud.[145]

The native range of *L. dubia*, the longnose spider crab, extends from Cape Cod to southern Texas, down to the Bahamas and Cuba, so these crabs are at the northern end of their range.[146] These crabs are also known as doubtful spider crabs ("*dubia*" from the same origin as dubious) and

I would love to know why. Both spider crab species have been called "decorator crabs" for the tendency to attach bits of seaweed, like sea lettuce to their shells. Using hooked, Velcro-like setae on the surface of the carapace, the crabs attach bits of algae and invertebrates for camouflage. This behavior is most common in juveniles which need the extra protection. The shells of adult crabs are usually found clean because they are now too large to fit in most predators' mouths. Under the decorative covering, the carapace of *L. dubia*, has six spines down either side, and along the median line on the dorsal surface.[147] One distinguishing feature that separates the long nosed and the portly spider crab is the number of spines, *L. emarginata* has nine spines. A forked rostrum extends between the eyes for the longnose spider crab and this looks a bit like a horn, and the overall color of the body is yellowish-brown.

Although the longnose spider crab is primarily a benthic species, it has been associated with several pelagic (mid-level swimming) organisms, including, the loggerhead sea turtle, the cannonball jelly, the sea nettle, the sea wasp, and the moon jelly. In fact, juvenile *L. dubia* have a special commensal relationship with cannonball jellyfish (*Stomolophus meleagris*). They literally hang out with cannonball jellyfish, sometimes inside the bell, for protection, for a free ride, to eat some of the food the jellyfish "drops," and they may even nibble on the jellyfish itself.[148]

I do not know why, but spider crabs always remind me of sheep in a tank because they tend to be gentle versus the much more aggressive blue and lady crabs. They also are scavengers and algae eaters although they have been shown to eat the tissue of live jellyfish.

Crabs have inspired many field station student research projects who have investigated the self-decorating habits of spider crabs, the swimming rituals of blue crabs, and the habitat preference of lady crabs (sandy versus silty bottoms) which is the subject of the next essay.

The Beautiful and Vicious Lady Crab
(August 27, 2015)

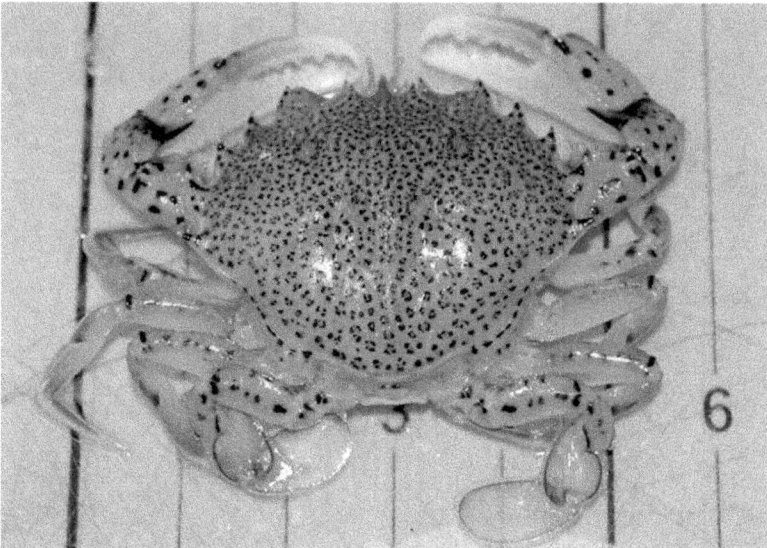

Photo by the author, UMass Boston Nantucket Field Station archives

One of the most charmingly aggressive creatures you might encounter on Nantucket is the "lady crab" whom you will find scuttling around on beaches and in marsh inlets. I had never heard of them before I moved to Nantucket and have always been fascinated by the purple "leopard print" markings adorning their shells.

Lady crabs tend to be extremely aggressive, even for crabs. They are basically the "James Bond woman" of crabs. Their scientific name is *Ovalipes ocellatus* and they are found in the waters off eastern North America. They are also known as leopard crabs, calico crabs, or ocellated crabs; ocellated means to have "eye-like markings of spots" and true to its name lady crabs have a shell covered in clusters of purple spots. They can be found all the way up in Canada and down to the shores of Georgia, and their favorite food is mainly mollusks, such as juvenile Atlantic surf clams.[149] In their profile online they claim to enjoy long walks on the beach.

The lady crab was first described by Johann Friedrich Wilhelm Herbst in 1799 who named it *Cancer ocellatus*.[150] In 1898, Mary Jane Rathbun moved the species to her new genus *Ovalipes* during her discoveries described in the "The Brachyura collected by the U. S. Fish Commission steamer Albatross on the voyage from Norfolk, Virginia, to San

Francisco, California, 1887-1888".[151] I bet that was a fun journey.

The lady crab is not only brightly colored but a true swimming crab. In the water and under direct sunlight, this crab's coloring appears iridescent. The species is called the lady crab because of the beautiful color patterns on the shell, obviously there are male and female lady crabs or this would be a very short article. The spots have a similarity to leopard spots or even appear as if someone put on purple lipstick and then kissed the crab, at least that is my story, and I am sticking to it. When you see sun-bleached molts or dead crabs on the beach, you'll notice that the bright purple color has faded to a pink or coral tone.

The carapace of *O. ocellatus* is slightly wider than long; full grown ones will be around 8.9 cm (3.5 in) wide, and 7.5 cm (3.0 in) long. Lady crabs have very sharp, powerful pinchers which are whitish in color with purple-spotted tips and jagged "teeth." The last pair of legs are modified into paddles and are adapted for swimming. Three sharp points are present between the eye sockets of the lady crab, as well as five sharp points along the carapace that turn toward the eye sockets. The number of points along the carapace helps to distinguish this crab from other similar crabs.[152] Like many other crabs, the females have large, rounded abdomens, this shape is not just coincidental, the female carries the eggs under a flap on her abdomen and when the eggs hatch, they are planktonic, simply drifting with the current. Young crabs hatch in the early summer months during which they become food for many fish. As they grow, they pass through two main stages (five stages overall) called the zoea and megalopa before becoming an adult and settling down into the benthos (bottom) usually in early fall. *O. ocellatus* goes through a grand total of five larval stages, lasting a total of 18 days at 25 °C (77 °F) and a salinity of 30‰, and 26 days at 20 °C (68 °F) and 30‰.[153] The warmer the water the faster the process. Nantucket Harbor is usually around 30-31.5 ‰ and for the latter part of the summer, around 25 °C or warmer.

The meat of lady crabs is not considered as tasty as that of other crabs, so they are not harvested commercially. Unfortunately, the fact that they are not a commercial species means that not nearly as much information available on lady crabs versus the enormously valuable and tasty blue crabs.

Lady crabs make up for their second-class status with their aggressive disposition and sharp claws. Part of the aggression stems from the fact that like other swimming crabs, the lady crab does not have a very thick or rigid carapace. This means it has less protection than some other crabs, which have harder shells or exoskeletons. Because it is not very well protected, it makes up for this by its speed, feistiness, and camouflage. Lady crabs are often seen partially buried in sand with only their eyestalks protruding. If you go snorkeling in Nantucket Harbor that is precisely what you will see. This species of crab is most often found in sandy substrates in quite shallow water, i.e., the surf zone. Normally, this is a very difficult habitat, because of strong wave action and constantly shifting sands. But the lady crab stabilizes itself and allows itself to ambush more food by burrowing just beneath the sand surface. As waves toss the sand around, the crab quickly shifts position and digs back under the surface. The lady crab will dart out of its hiding place using its powerful paddles to swim after its prey. Like most other crabs, lady crabs are scavengers, eating both dead and live fish, crabs, and other invertebrates. Lady crabs are on the dinner menu for oyster toadfish, tautog, striped bass, American lobsters, whelk, and other crabs.[154]

Ecologically, almost all crabs function as scavengers, feeding on dead or dying animals or organic debris of any kind. They are, in fact, like the vultures or hyenas of the sea. But they don't sit around waiting for things to die. Many species including the lady crab are also active predators, capable of killing small fish and breaking open the shells of different mollusks. Lady crabs are somewhat messy eaters, ripping apart their prey with their pincers and sweeping the bits into their mouth using a series of appendages called maxillipeds and maxillae. A pair of comb-like mandibles guard the mouth and chop the food into tiny pieces, which are then swallowed. A short esophagus leads to a gastric mill analogous to the gizzard of other animals, where food is ground up.[155]

All crabs have exoskeletons, which mean they must molt to get larger. These molts often confuse people as they appear to be dead crabs. These crabs appear to stop molting when they reach a width of four inches. Adults that have stopped molting may be covered in growths of barnacles or seaweed. Lady crabs prefer sandy bottoms and higher salinity parts of estuaries versus other crabs that can be found in muddy bottom areas. Look for them next time you are out swimming in the harbor. We will often have huge swaths of dead lady crabs eaten by seagulls that can be found in the wrack line entangled in the eelgrass in what I call the "lady crab apocalypse" in early summer. For several summers (2009-2016) interns monitored the population of lady crabs relation to blue crabs and other native and nonnative crabs. Blue crabs are less plentiful, but the students did mark and recapture several very large ones in Folger's Marsh. Although blue crabs get all the press for being aggressive crabs, Eugene H. Kaplan noted that lady crabs are the worse when it comes to vicious behavior and said they were "the crab most likely to pinch a wader's toes".[156] I would have to agree.

Beach Houdinis: Mole Crabs
(May 21-27, 2009)

One of the things I enjoyed the most as Director of the Nantucket Field Station was introducing high school students from cities such as Worcester, Brockton and Lawrence to some of Nantucket's flora and fauna as we explored various habitats around the island. One of their favorite activities was a morning spent at Codfish Park in 'Sconset. We took samples of the sand to determine average grain sizes using a stack of sieves back in the lab and measured the change in elevation from the back dunes to the rapidly growing midsection of the beach down into the surf zone. We also talked about how creatures adapt to the pounding waves, saltwater immersion, intermittent desiccation, and constant pressure from inquisitive predators like sanderlings (cute but hungry). We found an excellent example of a critter that exhibits all these characteristics and is one of Nantucket's literally hidden treasures when we came upon thousands of tiny tunneling creatures called "mole crabs."

The intertidal area of our beaches can seem to be relatively lifeless to the casual observer. In the swash zone where surf pounds the sand and waves run up the beach, there is a sea of life hidden under the sand that is the basis for an intricate food web. If you scoop up some sand in the swash zone (wet area on a beach), you'll see a flurry of little bodies burrowing down, scattering in all directions. You are watching the frantic stampede of the mole crab, *Emerita talpoida*, also known as the Atlantic sand crab to distinguish it from its Pacific brethren. An Italian naturalist, medical doctor, and expert on mercury poisoning named Giovanni Antonio Scopoli first described the *Emerita* (Latin feminine form for retired professor, bishop, or professional) genus which includes five other sand crab species in his 1777 work *Introductio ad Historiam Naturalem*.[157] The American naturalist Thomas Say officially laid claim to the discovery of the crustacean and subsequent taxonomic flag planting by selecting the descriptive species name "talpoida" which is derived from the Latin root "talpus" and Classical Latin word "talpa " for "mole."[158] It is pretty obvious how they got their name if you watch them quickly bury themselves into the sand when exposed. Mole crabs have several other common names such as sand crabs, sand fleas (which is a bit confusing because true sand fleas are tiny amphipods), sand fiddlers, and beach hoppers. If you've ever visited a beach on the East Coast, you most likely have encountered mole crabs without even knowing it. Perfectly camouflaged, if exposed by a wave, a person's foot or a child digging at the surf's edge, it will dive back under the sand before you even know it's there.

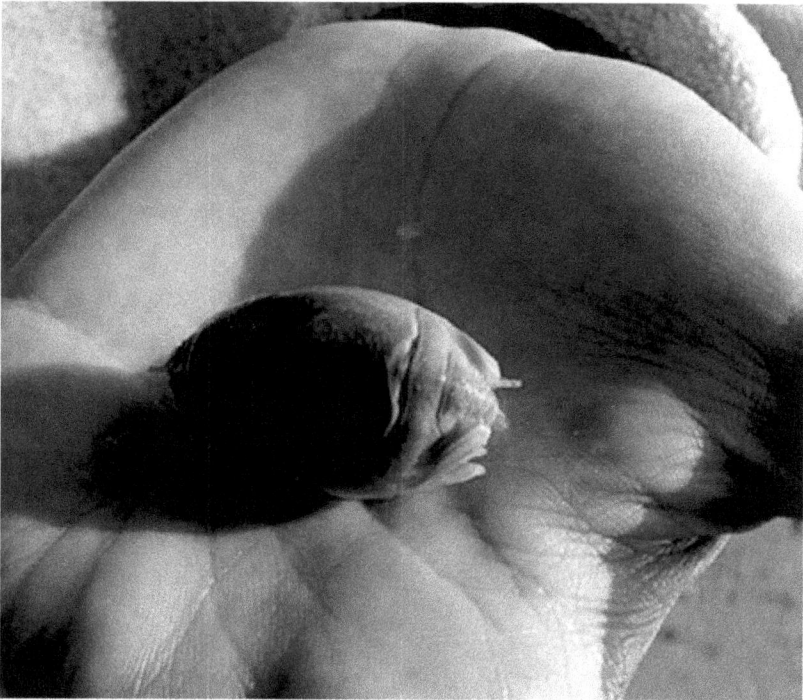

Photo from the author's archives.

This crustacean has evolved to have a body uniquely suited to thrive in the surf zone. They range in size from 1/4 of an inch to over an inch. The females are larger than the males. Their lifespan is two to three years. They are light pinkish/sand colored and live mostly buried in the sand in the intertidal zone. These little one-inch creatures have ten legs, like other crabs, but these are adapted for swimming and digging, not walking or defense. Mole crabs are found along the Atlantic coast from Massachusetts southward. The mole crab is an egg-shaped crustacean with a smooth rounded carapace. Not exhibiting a typical crab shape, this crab's abdomen is broad in the front and tapers to its tail, which has a pair of forked, leaflike appendages. A long, spear-like tail piece folds under the body and is used for anchoring in the sand and protecting the eggs as we'll see later. A pair of ball bearing like black eyes sit atop long, thin eyestalks. Unlike other crabs who tend to move side to side, mole crabs dig and swim backwards, using the flattened legs that are tucked into the carapace equally well as either oars or shovels. Nearly everything about its anatomy has evolved to facilitate getting under the sand as quickly as possible. They are known as "swash riders" and from now on I may call them "wave cowboys" because they move up and down with the tide to stay in wet sand and somehow avoid getting flipped over by wave action.

This species has two sets of antennae. The first pair are hairy and used as sensors to pick up vibrations from approaching waves. They also form siphons to draw in water to the gills when the animal is buried. The larger feeding antennae resemble an old-fashioned feather pen. The hundreds of bristles on the antennae are extended out into the water as a wave passes over them. These trap the plankton and then the little front legs are used to wipe off the food and transfer it into the mouth. The crabs always bury themselves in the sand facing the ocean, with their backs to the beach. The breathing antennae are closed together to form a small funnel that takes in oxygenated water and filters sand grains away from its gills. Sometimes you can see the "V" shaped ridges they leave in the wet sand as they filter the water, or you may even see tiny bubbles marking the spots where they are hiding.

Between tides, they dig into the sand to hide from the shorebirds who would eat them (larger crabs and fish also make meals of them). Only their eyes and antennae peep out above the sand. The feathery feeding antennae are also used for cleaning themselves, truly an "all-purpose" appendage suitable for late night infomercial fame. There are a lot of fascinating You Tube videos showing the feeding behavior and "now you see them, now you don't" disappearing act of the mole crabs. The effectiveness of the feathery antennae for filter planktonic bits from the water column is especially obvious in some of the videos.

Females grow to about one inch, while males grow to about half an inch. Males reach sexual maturity when they are only 0.125 inches in length. Both sexes move into deeper water in the winter to avoid temperature extremes. In summer, orange egg masses will be noticeable on the abdomens of the females. While it is not easy to know the first time which end of a mole crab is the front, you can find out by holding one in your hand and watching which direction the crab moves: the end that moves first is the rear end. When placed back into the sand, they can bury themselves in seconds. They don't have pincers or claws, so they can't hurt you, which is why children enjoy playing with them. Do be kind and put them back where you found them, preferably on their bellies so they can escape.

Like most crabs, *Emerita* is a brooder, which means they carry the eggs on their body and protect them until they are ready to be dispersed, not that they are emo. The female holds up to 50,000 eggs in the swimmerets beneath her tail. They are protected there until they hatch, and swarms of larvae are carried away by the tide. The female times each of the egg releases to coincide with the outgoing waves so that the larvae aren't stuck "high and dry." To prevent the larvae from washing up on shore immediately after hatching, the mother will tuck the tail against her body as a wave rolls in, then extend the tail, releasing the young as the wave recedes. This is done multiple times until all the larvae are released. Because the planktonic young may move well offshore before settling to the bottom, they must swim a long distance to find their way to the nearest surf zone. Very few of them survive this trek.

This species is commonly used for bait for flounders, drum, and other sport and commercial fish and they are collected from the intertidal area by fishermen using wire mesh rakes. Some people even eat them, although it takes several pounds to make a meal. Mole crabs are an important, some might say even critical, food source for shorebirds who forage in the surf zone. They have been used recently to evaluate pollution effects and other stressors on beach habitats. Mole crabs have also been used around the country as indicator species, or "canaries in the coal mine" to evaluate beach nourishment effects on intertidal areas and to measure the concentrations of toxins from harmful algal blooms. Recently, scientists studying the Pacific coast mole crab species *Emerita analoga* have found that microplastics are causing mortality and reducing their ability to hold onto their egg clutches.[159]

Other scientists look at the distribution of various age and size classes (juvenile, immature and mature male and female) mole crabs in the intertidal area to see how quickly these creatures can move and to figure out how they manage to stay in the surf zone, essentially, they are studying mole crab dance patterns.[160] The adaptability of these creatures and their toughness may win out over threats to their existence and ensure that they can support the food chain.

Next time you are at any Nantucket beaches, dig in the upper sections of the surf zone and see if you can find this hidden treasure.

Ghost Crabs
(August 1, 2013)

The name of this essay sounds like an awesome indie horror film, doesn't it? Well, this story tells the tale of what a keen eye and good observation skills can do. Thanks to the eagle eyes of a former Manomet Center for Conservation Studies shorebird monitor (and one of my favorite people) Edie Ray, ghost crab holes and tracks were found at Smith's Point in 2013, the first officially documented sighting on Nantucket.

What is a ghost crab? How might they change the beach habitat that they share with mole crabs and sand fleas? Why are they appearing on Nantucket?

Ghost crabs, also called sand crabs, are semiterrestrial crabs of the subfamily Ocypodinae, which can be found all over the world in subtropical and tropical areas.[161] Male ghost crabs are slightly larger, and they have one claw that is bigger than the other claw, but this difference is not as noticeable as it is in male fiddler crabs (who sometimes have humongous claws for their size). This trait is called sexual dimorphism, in which the physical traits between male and female members of a species vary in size or shape. Worldwide, there are roughly 20 species of ghost crabs, but *Ocypode quadrata* is the only one found on the east coast of the United States. The species is a small sand-colored or grayish-white crab, measuring at most a little over 5 cm (2 inches) across the back at maturity.

Atlantic Ghost Crabs have squarish sand-colored shells having margins (edges) that are finely beaded but toothless; the claws are white. The space between the eyes is much shorter than the eyestalks.[162] They really look comical, almost like a cartoon crab. Ghost crabs are some of the most terrestrially adapted of crabs, they only must have access to salt water for two things: females release their eggs at the water's edge (larvae are aquatic) and they moisten their gills either in the surf or from wet sand.[163] Ghost crabs belong to the Order "Decapoda" which means ten legged. I bet when you have looked at a crab you haven't thought, look at all those legs?! That's because some of these appendages have been adapted into mouthparts and claws.

The name ghost crab alludes to their pale ghostly color, daytime elusiveness, and their penchant for nighttime activity (nocturnal creatures).[164] The scientific name Ocypode is derived from the Greek roots "ocy" meaning "fast" and "podos" for "foot", in reference to the animal's speed.[165] And they are indeed super-fast. They can run up to 10 miles an hour and change course on a dime. When I lived in Galveston and would walk the beach at night, you would see virtual armies of these guys popping in and out of holes. It was very eerie. The species name comes from the Latin "quadratus" for "square" referring to their squarish carapace or body. The ghost crabs come out at night to feed (sounds like a zombie movie doesn't it?) to avoid predation by shorebirds like gulls.

Its range of distribution extends from its northernmost reach on Rhode Island's beaches south along the coasts of the tropical Western Atlantic Ocean to the beach of Barra do Chui, in Rio Grande do Sul in Southern Brazil. They are found on the supralittoral zone (the area above the spring high tide line) of sand beaches, from the water line up to the dunes.[166] They have moved onto the southern shores of Block Island and Martha's Vineyard for the past 3-4 decades and are now being found on our shores.

Their stalked compound eyes can swivel to give them 360° vision. Young crabs are cryptically colored to blend in with their sandy habitat. The young crabs build their holes closer to the water while older crabs build theirs further away from the high tide line, sometimes as far as a quarter mile from the tide line. The burrow down as far as 4 feet and can close the top of their burrows with sand when it is very hot or very cold. Their gills allow them to carry of pillow of oxygen close to their body which they will use to sustain themselves when they hibernate over the winter. Juvenile and females make their holes sloppily, but adult males will make a nice pile of sand balls very similar to the pseudo feces that fiddler crabs adorn their hole entrances with (look honey, I am a great mate, my lawn is so organized!).[167] Adult ghost crabs dig deep burrows, comprising a long shaft with a chamber at the end, occasionally with a second entrance shaft. They remain in the burrow during the hottest part of the day, and throughout the coldest part of the winter. According to Wayne and Martha McAlister's book *Life on*

Matagorda Island: "Ghost Crab tunnels enter the sand at 45° angles and are directed so that they catch the onshore breeze. Often two tunnels join into a Y, possibly to create a draft through the lair. Other tunnels are simple Ls or Js, the bottoms enlarged so that the crab can rest and turn around during the day when normally it's belowground. The tunnels descend to just above the water table, and usually there's just one crab per tunnel."[168]

Ghost crabs, in keeping with their names, emerge mostly at night, to feed on mole crabs (*Emerita talpoida*), clams, and other beach crustaceans. They are omnivores who will eat a wide range of items, including carrion, plant material, debris and occasionally turtle hatchlings further south. Their presence indicates that a beach is in good health. *O. quadrata* can produce a variety of sounds, by striking the ground with the claw, by stridulation with the legs (more about that below), and sometimes emitting an incompletely explained "bubbling sound." Some people say they can also hear their digging noises underground from as far as six feet away.

Ghost crabs have a ritualized competition for a female ghost crab's affection, which involves a choreographed fight between two males, raising both their claws as well as their bodies in intimidating poses until one sinks into a submissive posture and gives in. If this doesn't happen right away, a "pushing fight" ensues until one crab finally withdraws. All this hullaballoo happens at night. Ghost crabs have a unique mechanism on its right claw known as a stridulating organ. When it strokes this against the bottom of its leg, a squeaky noise is produced. A crab produces this noise to warn other crabs not to enter its burrow. Male crabs also use this sound to attract female mates so they can have something to fight over. Usually, crabs tend to reproduce continuously in tropical and subtropical regions, because of the favorable environmental conditions for availability of food, development of gonads, and liberation of larvae, whereas in temperate zones reproduction is restricted to the warm months. The females release their eggs in the surf zone (if held underwater ghost crabs suffocate so this is more dangerous than it looks). The larvae take about six weeks to develop in the ocean currents before they wash back ashore.

It appears that scientists are just now starting to undertake more serious population surveys of the creatures further south on the Atlantic coast where they are more common. As ghost crabs move north, they will likely replace sand-hoppers which are the more dominant beach organisms in cooler areas like New England. It is somewhat hard to find out what eats ghost crabs; in some locations, hawks, shorebirds, and gulls are their primary predator. Scientist found that ghost crabs were a significant part of the diet of the burrowing owl *Athene cunicularia* in

Brazil.[169] Their nocturnal habits are thought to be their way of avoiding becoming a snack for shorebirds.

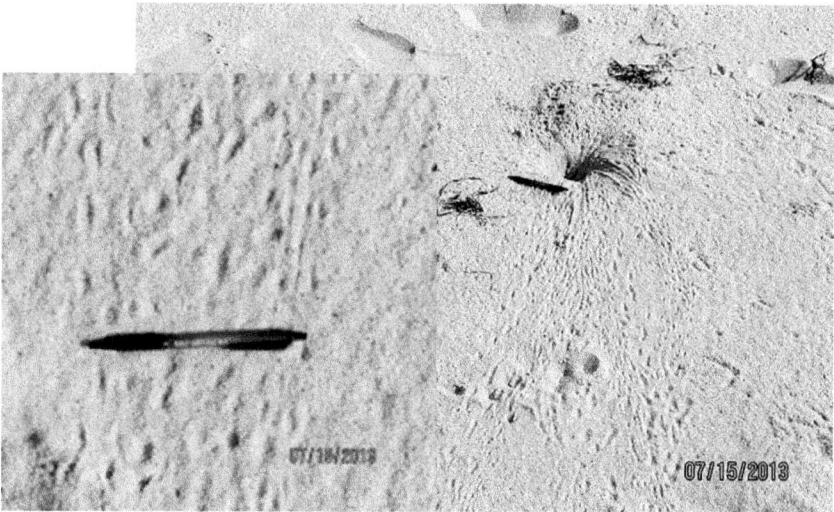

Photos by "Edith" Edie Ray, provided to the author.

Ghost crabs have been seen on and off on Martha's Vineyard for years despite being listed as only appearing as far north as Long Island. The aquatic larvae probably washed up on our southern shores from eddies and other offshoots of the Gulf Stream.

Luanne Johnson, a Martha's Vineyard biologist completing her PhD with Antioch University and the Director of the nonprofit organization BiodiversityWorks has been tracking these for the several years on our sister island. Other MV associated biologists such as Dr. Tim Simmons, with the Massachusetts Natural Heritage program saw small ones on Martha's Vineyard back in the 1980s but has been noticing them getting bigger and bigger.

And island naturalist, former director of the Felix Neck Wildlife Sanctuary, and current owner of the World of Reptiles in Edgartown, Gus Ben David, has seen them for close to 40 years.[170] Whether this is an offshoot of climate change or simply a normal extension of range, it is a bit early to tell. One thing that may be worrisome to scientists on both islands is the predilection of these crabs to eat plover eggs and chase chicks. The last thing piping plovers need is an additional predator that they don't expect (something like the Spanish Inquisition!)[171]

In a recent study conducted in North Carolina, Peterson et al. (2000) found significant deleterious effects on ghost crab populations resulting from beach nourishment and bulldozing on eroding beaches. In addition to these practices, the loss of intertidal beaches to shorefront armoring was also considered a significant threat to ghost crab populations in South Carolina. Not only does this compromise the ghost crabs themselves, but these activities wipe out their primary food sources.[172]

The large burrows Edie found and that have been found on the Vineyard indicate that these bad Larrys overwinter on island. That is amazing considering how tough recent winters have been. As our world warms, more southern species will migrate north. Next time you are at the beach listen for their digging or look for their holes and imagine their nightly marches.

Neighborhood Bullies, the Green Crab
(July 24-30, 2008)

Imagine yourself as a typical Nantucket Bay scallop, minding your own business in the shallow water on a blade of eelgrass. Before you can squirt away, a green crab sidles up and attacks! This invader, whose Latin name is *Carcinus maenas*, is an invasive species in North American waters and was transported from Europe in the 1800s to disrupt the ecological balance of our harbor. Common names for this small crab that quickly became a big threat include the European green crab, shore crab, and the New England nickname used by fishermen in the 1900's of "Joe rocker." *C. maenas* was first observed on the east coast of North America in Massachusetts in 1817.[173] It was first introduced to the Pacific coast in San Francisco Bay in 1989.[174] Since its introduction along the Pacific coast, it has traveled over 500 miles in 10 years (from Monterey Bay to Vancouver Island)[175], which is pretty darn fast for an aquatic creature. The European green crab is listed as number 5 on the top 10 list of "Animals Least Wanted" by the Nature Channel with other alarming interlopers like fire ants, European starlings, and brown tree snakes.[176] Today, the green crab has the distinction of being one of the most prolific crabs in the New England area.

Carcinus maenas is a small shore crab (adults measure about 3" across) whose native distribution is along the coasts of the North and Baltic Seas. Although known by the common name of "green crab," the carapace (top shell) color can range from green to gray to yellow or brown in a mélange of camo-like slimming shades. Juveniles can change color to match their surroundings. Adults are generally dark greenish with yellow markings. The underside is often bright red or yellow. Green crabs do not have the last set of swimming legs or paddles seen on the rear of blue crabs, although their last pair of legs are slightly flattened and lined with little hairs or setae. Green crabs can be right or left-handed, and for left-handed crabs, and especially males, the claw (cheliped) size is larger (sexual dimorphism) although the claw size differential is not quite as pronounced as it can be for other species. They can also be identified by the five short teeth along the rim behind each eye, and three undulations between the eyes.[177] If you are looking at a live one that close, by now you'll have a claw in the eye as they are very aggressive.

What makes this invasive species so prolific and dangerous? The green crab is an effective forager, adept at opening bivalve shells. Studies have shown it to be quicker and more dexterous than other crabs and capable of improving its food gathering skills over time. It preys on a multitude of organisms, including clams, oysters, mussels, marine worms, and small crustaceans, making it a major potential competitor of our native fish and bird species. At the turn of the century, this species basically wiped out the soft-shelled clam industry of Maine and the surrounding waterways, and it has been implicated in shellfish population reductions for both bay scallops and quahogs.[178] It has been documented that bay scallops can experience up to 70% mortality due to green crab predation, although they were observed to be able to fend off the green crabs more successfully once they reached a decent size toward the end of their first year of age.[178]

C. maenas can live in all types of protected and semi-protected marine and estuarine habitats, including habitats with mud, sand, or rock substrates, submerged aquatic vegetation, and emergent marsh, although soft bottoms are preferred. *C. maenas* is euryhaline, meaning that it can tolerate a wide range of salinities (from 4 to 52 ‰) and it can survive in temperatures from 0°C to 30°C.[179] The wide salinity range allows *C. maenas* to survive in the lower salinities found in estuaries. Apparently, it enjoys opening a new niche in the predator-prey equation in its adoptive areas as it grows larger in invaded habitat versus its natural habitat.

Green crab larvae can survive as plankton up to 80 days. Ocean currents disperse the larvae many miles up and down the coast. After a period of growth and development in the open sea, green crabs in final larval stage aggregate at night in surface waters. Tides and currents sweep them back into coastal waters where they molt and settle out as juvenile crabs in the upper intertidal zone. If the conditions in their new home are suitable, the crabs may survive and even reproduce, establishing a new population and extending the species' range farther along the coast. Some scientists believe they can even disperse by ocean currents due to El Nino and La Nina events, which have been known to move the larvae of organisms on a global scale. To add insult to injury, research in *Nature* by Kevin Lafferty, a USGS marine ecologist at the Western Ecological Research Center in Santa Barbara, California and his colleagues found that green crabs have much fewer parasites in their adopted homes versus their native locales.[180]

There are several ways humans can inadvertently disperse green crab to new habitats. Scientists believe that one likely pathway of introduction is through the distribution of live seafood. Green crabs are sometimes present in seaweeds packed with lobsters and commercial oysters. If the packing material and containers are not disposed of properly, the crabs can find their way into waterways. Although heavily regulated, the aquaculture industry is also a potential source of green crab introductions. Recreational boaters transport nuisance species in bait buckets or boat wells, often without realizing it. Live green crabs are also used as bait by recreational fishers or are present in the seaweed packed with bait. In addition, they are available for purchase from marine biological supply companies.

Scientists have also identified ballast water as a major pathway for aquatic introductions, including the larval stage of green crab. Marine vessels take on and discharge millions of tons of water for ballast each day, which may contain aquatic plants, animals, and pathogens. When a vessel unloads or picks up cargo, the operator often empties the ballast tanks, thus introducing a myriad of marine life from bacteria to adult fish. In addition, barnacles, mussels, seaweed, and an abundance of other marine life attach to hulls, rudders, propellers and piping systems. Once in port, the organisms can reproduce or become dislodged and swim away.

The United Nations International Maritime Organization (IMO) instituted voluntary guidelines in 1991 calling for ships to release their ballast water on the high seas and refill the tanks with mid-ocean water, based on the assumption that species from coastal zones will not survive in the open ocean.[181] Both Canada and the United States have established voluntary guidelines for ballast water exchange and require that all vessels entering their 200-mile Exclusive Economic Zones (EEZ) file a ballast water management report. Unfortunately, compliance, particularly in the New England area, can be spotty. Experts on the transportation of marine aquatic invasive species are investigating the use of a combination of ballast treatment techniques such as ultraviolet light (UV), ultrasound, and ozone sterilization to eradicate the miniature stowaways before they are dumped in our harbors.[182]

Once it arrives, the green crab can thrive in many types of coastal habitats and in wide ranges of temperature and salinity. The green crab can produce up to 200,000 eggs at a time, and under certain conditions, it can survive up to two months out of water.

Although green grabs are listed as an edible species, for many, picking out the small amount of meat was considered too tedious, as a result, the crab was thought to have little commercial value. But scientists and marketers and innovators are all working on creating a demand for this introduced species. In the last 5 years and especially in 2021 and 2022, various groups have been marketing green crabs. You can eat them on the soft shell at Row 34 restaurant in Boston or try the whiskey made from a pound or so of crabs called "Crab Trapper," People say it is like a spicy fireball.[183] Scientists have also been extracting a variety of proteins to evaluate their efficacy in treating disease such as Type 2 Diebetes.[184] This is the ultimate version of making lemonade out of lemons!

In 1995, in Edgartown, Massachusetts a bounty was paid for green crabs as part of a response to the threat to commercial shellfish. Approximately 10 metric tons of crabs were trapped in the local salt ponds, which was presumed to improve survivorship of hatchery-reared scallops and hardshell clams.[185, 186] This approach was also identified as an action item in the Nantucket and Madaket Harbors Plan. The most successful bounty program has come from the Northern Gulf of Maine launched by the Parker, Ipswich, and Essex River Watershed Collaborative (PIE). Initiated in 2013 by PIE-Rivers partner MassBays and the "Eight Towns and the Great Marsh" program, the Towns of Ipswich, Rowley, Essex, Newbury, and Gloucester have engaged in the trapping of green crabs based on a $0.40 per pound bounty funded by the MA Division of Marine Fisheries and the Town of Ipswich. To date, more than 100,000 lbs. of green crabs have been removed from the waterways.[187]

Fortunately, for the Chesapeake Bay area's lucrative blue crab (*Callinectes sapidus*) fishery, green crabs are not found in Chesapeake Bay, but they are at the mouth of the Bay. Scientists find low concentrations of green crabs where blue crabs are which indicates that the blue crab could be used as a control agent for green crabs.[188] But, other researchers have found that when juvenile blue, green and shore crabs were put together in a tank, the green and asian shore crabs were more antagonistic than the blue and the blue's carapace was more prone to breakage (i.e. it lost more of the fights and had to expend the most energy).[189] Some scientists have even investigated a biological control organism, the parasitic barnacle *Sacculina carcini*, but initial tests show that this barnacle orefers green crabs substantially but is not host specific and would infiltrate other crabs such as Dungeness crabs.[190]

But there is another relatively recent arrival to our shores that may cramp the green crab's style. The invasive Asian shore crab (*Hemigrapsus sanguineus*) was discovered in New Jersey in 1988. Marina floats towed from New Jersey to Massachusetts in 2000 brought with them a population of Asian shore crabs. Measuring two to three inches wide, they have spread as far north as Maine and as far south as North Carolina. Like the green crab, this Japanese import reproduces in far greater numbers than native species. The crab dines on worms, barnacles, shellfish, and algae. It has even been known to eat green crabs. Miraculously, it may leave most native species alone.[191]

In 1998, the European green crab was formally recognized as an Aquatic Nuisance Species (ANS) by the Federal ANS Task Force (a national coordinating body). In Washington State, they have a large volunteer early detection and trapping program designed to quickly verify first sightings of the invasive crab in various harbors and estuaries. It is illegal to possess a green crab in Washington State and they are also prohibited in Oregon and California.[192]

On Nantucket, proactive actions have been implemented to reduce the green crab population and protect shellfish populations. The shellfish biologists have been placing crab traps throughout the harbor to collect, quantify, then kill any green crabs found for several years. At the Nantucket Field station each summer, population estimates of crabs in Folgers' Marsh are determined through catch, mark, and recapture experiments. Parts of the marsh with more blue crabs have fewer green crabs and vice versa. This is going to be an ongoing issue very similar to all the other invasive species explosion occurring on waterways and in harbors around the country.

Stealing Home
(August 27-Sept 2, 2009)

One of my favorite creatures to house in the Field Station aquarium were hermit crabs. They fascinated children with their antics, "musical shell" games, and obsessive noshing on floating detritus while simultaneously terrorizing some of the smaller fish. Hermit crabs are not only tank entertainers in marine aquariums, but they are also very common household pets. Every parent at one time or another has been the reluctant guardian of either a terrestrial or aquatic hermit crab. My nephews had two land hermit crabs that lived for seven years, long after the excitement had worn off. In fact, if properly cared for, some hermit crabs found at pet stores or at county fairs can live for decades.

Hermit crabs are in the Decapoda ("ten footed") order, the infraorder Anomura (which includes mole crabs and sand crabs) and the Superfamily Paguroidea (Latreille, 1802). The members of Paguroidea have oval carapaces which are usually asymmetrical.[193] They live either in shells (hermit crabs) or with their abdomen tucked underneath (stone, coconut, and king crabs). These latter crabs like the king crabs which have abandoned their shell home for a free-living life are called "carcinized" crabs and are thought to be more similar to true crabs. In evolutionary biology, carcinization is a hypothesized process whereby a crustacean evolves into a crab-like form from a non-crab-like form (can be observed on Nantucket in humans in the winter). The term was introduced by L. A. Borradaile, who described it as "one of the many attempts of Nature to evolve a crab."[194] The hermit crab morphology reveals they are not closely related to true crabs and instead share more characteristics with crustaceans like lobsters.

The online Etymology Dictionary traces the term "hermit" from the old French word "(h)eremite" which further descended from the Late Latin word "ermita" and Greek word "eremites" which means "person of the desert," which derived originally from eremia "desert, solitude," and "eremos" for "uninhabited."[195] It is easy to see why hermit crabs are called that considering the majority of their lives are spent isolated in a single room or cell like a mythical hermit living alone in a small cave.

Photos of hermit crabs at the Nantucket Field Station courtesy of Mirabai Perfas

Most hermit crabs are anything but hermits and prefer to live in groups and have developed foraging techniques that depend on group dynamics and cooperation for feeding success and for protection from predators. Because of their gregarious nature, hermit crabs will survive as pets longer when paired with a buddy. Once thought to be throwaway animals, hermit crabs like the Caribbean hermit crab (*Coenobita clypeatus*) can live for 20-30 years if properly cared for.[196]

An early ancestor of the hermit crab may have existed 500 million years ago as theorized by geologists James W.

Hagadorn and Adolf Seilacher. They believe that the adaptation of coiling an asymmetrical body into a snail shell was a viable method to migrate a water breathing creature onto dry land to expand habitat and food sources. They found evidence of their theory in fossilized shell "trails" preserved in ancient Cambrian intertidal sandstones recorded in a thin layer of microbial film. These early crab-like creatures were tucked into tiny shells which only provided them with moisture, essentially bringing their aquatic home with them.[197]

In our modern world, there are about eight hundred known species of hermit crabs, most of which are aquatic and live in saltwater at depths ranging from shallow coral reefs and shorelines to deep sea bottoms. However, in the tropics, several species are terrestrial, and some of these are quite large, like the Giant Hermit Crab (*Petrochirus diogenes*). Some can even climb trees like the Caribbean hermit crab (*Coenobita clypeatus*).[198]

Hermit crabs have evolved over time to be uniquely suited to their lifestyle. Their entire bodies have slowly conformed to fit into their shells. Unlike most other crab-like species, their bodies are asymmetrical so that they can curl into snail shells. Hermit crabs are commonly seen in the intertidal zone, for example in tide pools. Land crabs need dry land to survive and will die if they are in the water too long although they need some water to keep their body and lungs moist. The ocean crab can't be taken out of the water too long or it will die.

Large and small hermit crab "buddies" in the tank at the Nantucket Field Station. Photo by Mirabai Perfas.

Most species of hermit crabs have long soft abdomens which are protected from predators by the adaptation of carrying around a salvaged empty seashell into which the whole crab's body can retract. Hermit crabs prefer to use the shells of sea snails or marine gastropod mollusks such as our northern moon snail or whelks. The tip of the hermit crab's abdomen is adapted to clasp strongly onto the inside column of the snail shell. Their front claws have evolved to fill in the opening of a snail shell when the crab is retracted to form an operculum (Latin for "little lid" or door).

The crab life cycle is an involved one with several steps. In the late spring, you'll see hermit crabs traveling in the intertidal part of the beach in pairs with the males sometimes dragging the females around by their shells (hopefully more romantic than it sounds). After mating, the female keeps the eggs safe inside their shells until they are big enough to release. The first two stages of a hermit crab's life (the nauplius and protozoea) occur within the egg. When the eggs are ready, she crawls partway out of her shell to brush them out using a hind leg into the ocean, where they hatch into tiny, free-swimming larvae called zoea. Hermit crabs require this aquatic step to hatch their eggs which makes it very difficult to breed them in captivity. Zoea grow and molt several times before becoming megalops, which are still tiny but have a crustacean-like form. They develop their claws and antennae during these stages. Megalops molt into juvenile hermit crabs. Juveniles continue to grow and molt, eventually becoming adults.[199]

Every time the hermit crab grows, its exoskeleton doesn't, so it needs to molt and grow a new one. It then needs to find a new shell to live in that fits it. Once it finds a new home, unless it is checking out another one, it tends to stick close to its protection. It is extremely difficult to pull a hermit crab out of its shell, but you can drill a hole in the shell and poke it out (obviously not high on the fun scale for them). Hermit crabs may molt every two or three months when they are young and every 18 months as they get older. They can regenerate missing appendages such as their claws during the molting phase.[200] This regeneration is needed because hermit crabs occasionally self-amputate a claw if attacked, which is called autotomy.[201]

One fascinating fact about hermit crabs that relates to their social structure is their tendency to form "vacancy chains" or "housing chains," which I have seen in the shallow waters of the harbor. Recall, as they grow, they must find a larger shell to occupy. Several hermit crab species, both terrestrial and marine, have been observed forming a "vacancy chain" to order to exchange shells.[202] Typically, when a crab finds an empty shell, it will leave the safety of its own shell and inspect the vacant shell to see if it would be a better fit. If the shell is too large, the crab goes back to its own shell and then waits by the vacant shell for up to 8 hours. As new crabs arrive, this same practice occurs, they also inspect the shell and, if it is too big, wait with the others, forming a group of up to 20 individuals, holding onto each other in a line from the largest to the smallest crab. As soon as a crab arrives that approves of the size of the vacant shell and claims it, leaving its old shell vacant, all the crabs in line quickly exchange shells in sequence, each one moving up to the next size![203] But these (land based) hermit crabs aren't always patiently waiting. They use the socially created congo line that forms to "gang up" on an unlucky crab with a better shell, and pry its shell away from it before competing for it until the winner takes it over.[204]

Hermit crabs are omnivorous, which means they will eat anything from tiny plants and animals to decaying matter and detritus. Hermit crabs drink by dipping their claws in water then lifting out drops of water to their gills and mouth. They use their front claws almost like a fork and knife, holding larger food pieces with one claw and tearing them into bite sized morsels with the other. Hermit crabs are nocturnal animals which help them survive and reduce the likelihood of them drying out in the sun. They move around a lot more at night then during the day and some scientists have recorded them making "croaking sounds" which implies some amount of communication.[205] Some species have a mutualistic relationship with sea anemones that attach themselves on the shell, obtaining free transportation and scraps of food in return for protecting their hosts. Many of the hermit crabs we find here have barnacles or slipper shells on the outside of their host shells, getting a free ride while weighing down the crab.

Several hermit crab studies have been done at the Nantucket Field Station including a population census in 1988. On Nantucket we see three species on our beaches in the intertidal area: *Pagurus annulipes*, or the banded hermit crab which is the smallest one that favors periwinkle shells and has a hairy claw. The long-clawed hermit crab, *Pagurus longicarpus*, which the name suggests has long, narrow claws with a darker stripe on the "hands" of the claws. The flat-clawed hermit crab, *Pagurus pollicaris*, has a flat dominant claw with wart-like projections called tubercles. It is the largest hermit crab species found on Nantucket. For a couple of summers, we housed two huge ones in our tank living in large whelk shells. I usually would describe these hermit crabs as walking on their knuckles.

Dr. John Ebersole was a faculty member at the University of Massachusetts Boston, and he mentored graduate students conducting hermit crab science on island for more than twenty years. Back in 1995, one of his students, David Carlon, spent a few summers looking at the intricacies of the local hermit population and published several papers on the topic. In a nutshell (or hermit crab shell) their research showed that the little guys (*P. annulipes*) reproduce soon after metamorphosis and have a high reproductive effort while the other two species took longer to become parents and had fewer offspring. Size differences among species were related to patterns of shell usage. Male and female P. annulipes were always found in large shells relative to body size. In comparison, male and female P. longicarpus and P. pollicaris were found in small shells compared to body size. They discovered that the smallest hermit crabs were reproducing at a younger age and more frequently to compensate for their high risk of mortality associated with small shells. They also found that the shell needs (specifically large shells which are more rare) of the bigger species were a limiting factor for the population growth. Populations grew (or shrank) in concert with the number of available shells.[206]

This research supports other studies in which researchers found that if shells were provided for hermit crab populations, the populations grew (sounds like both fun research and a little weird). Several studies show that hermit crab shell use and exchange has become ritualized and competition for shells is quite fierce. In fact, most scientists believe and there is a fair amount of evidence to suggest that hermit crab populations including those of our three common ones on Nantucket are controlled by the number of available shells. In some parts of the world, hermit crab populations are declining because there are fewer and fewer shells for them to occupy, which sounds uncomfortably like the situation for humans around the world.

So, now we can think a bit about a very common trait that most of us have, which is the tendency to carry home a bucket full of shells from the beach. Occasionally I have thought about whether it would deprive a hermit crab of a future home, and some of my very young students have thought of this as well, but I bet you may not have. Next time you are on a beach and see a large whelk shell, think twice before carting it home, a nearby hermit crab may be eying it for a future renovation.

Counting "Living Fossils" & Dinosaurs Among Us
(June 11 - 17, 2009 and June 9-15, 2011)

In June of 2011, I was walking with a friend chest deep in the salty surf at high tide at 2:44 am. Did I lose a bet? Was I an extremely creative sleepwalker? Nope, just seeing how many horseshoe crabs were coming inshore to do what they have done for eons, which is to have crab relations, lay eggs, and continue the species. There's nothing more surreal than wading up to your chest in the pitch-black new moon darkness along the beachfront with headlamps and lanterns looking for silent, ancient, "living fossils."

Teams of volunteers from the various scientific organization on Nantucket have been monitoring horseshoe crab populations for many years by going out during the day and night at the times of the highest tides on the new and the full moon (and two days before and after) to count the number of horseshoe crabs present and record their activity. This ongoing (as of 2022) research is part of a multi-state effort to count the number of spawning horseshoe crabs along our shorelines to determine how the population is doing. Many years ago, Nantucket beaches, coves, and marshes were covered with horseshoe crabs. They were extremely abundant on the Cape and Island and throughout New England. Their numbers have declined over the past few decades due to a combination of factors, including overfishing, excess mortality occurring from harvesting them for their blood, and habitat loss.

Horseshoe crabs are sometimes referred to as "living fossils" which are defined as species who have not changed much from their ancient evolutionary roots and yet have very few close living relatives and few "branches" protruding from their phylogenetic tree. You might be surprised to find out how many plants and animals as diverse as the ginkgo tree, the koala, and the nautilus are classified as living fossils.

Photo by the author

No other creature has remained so close to its early form and shape. The horseshoe crab has been relatively unchanged since the Triassic period 230 million years ago, and similar species were present in the Devonian, 400 million years ago. Horseshoe crabs are the closest living relatives of the now extinct trilobites.[207]

Despite their common name, they are not crabs but are in the phylum Arthropoda (animals having an articulated body and limbs) which includes insects, arachnids, and crustaceans. Horseshoe crabs are in the own class called Merostomata or "legs attached to the mouth."[209]

The horseshoe crab was first named by Carl Linnaeus (1707 – 1778) a Swedish botanist, physician, and zoologist, who laid the foundations for the modern scheme of binomial nomenclature (two-part names that are often Latinized). Linnaeus is known as the father of modern taxonomy and is also considered one of the fathers of modern ecology. At the time of his death, he was widely renowned throughout Europe as one of the most acclaimed scientists of the time and his system for naming, ranking, and classifying organisms is still in wide use today.[209]

He called this creature *Limulus polyphemus*. The genus, "Limulus," from the Latin, meaning "somewhat oblique, odd, or askew" and referring to the sideways placement of the compound eyes and the species, "polyphemus," from the

Greek, meaning "one-eyed giant" perhaps a mistaken reference to eye spots.[210]

There are only four living species of the horseshoe crab family; our local buddies, *Limulus polyphemus*, and three species found in the Indo-Pacific. *L. polyphemus* is found along the western Atlantic and Gulf coasts from southern Maine to the Yucatan Peninsula, with the Delaware Bay as the center of the population. They are most abundant and larger in the middle of their distribution from Maryland to New York, smaller individuals and lower populations are found both north and south, most likely owing to less optimal temperature and salinity conditions.[211]

Horseshoe crabs have three main parts to the body: the head region, known as the "prosoma," the abdominal region or "opisthosoma" which is attached to the "head" by a hinge, and the spine-like tail or "telson."[212] It is the tail that earns this order its name Xiphosura, which derives from the Greek for "sword tail." The sexes are similar in appearance, but females are much larger than males. The carapace is shaped like a horseshoe and is greenish grey to dark brown in color. On the underside of the prosoma there are six paired appendages, the first of which (the chelicera) are used to pass food into the mouth. The second pair, the pedipalps, are used as walking legs; in males they are tipped with 'claspers' which are used during mating to hold onto the female's carapace. The remaining four pairs of appendages are the 'pusher legs', also used in locomotion. The opisthosoma bears a further six pairs of appendages; the first pair houses the genital pores, while the remaining five pairs are modified into flattened plates, known as book gills, that are used in breathing.[213]

Horseshoe crabs have several pairs of eyes. Two large compound eyes on the prosoma (head area) are sensitive to polarized light and can magnify sunlight 10 times. A pair of simple eyes on the forward side of the prosoma can sense ultraviolet light from the moon. In addition, multiple eye spots are located under the prosoma, with more on the underside of the tail. Horseshoe crabs occasionally swim upside down and may once have used these eyes more than they do today[214] (see evidence below upon finding out wader boots are not girlfriends).

Horseshoe crabs use their book gills (flaps resembling the pages of a book) to get oxygen from the water. If these primitive gills stay moist, horseshoe crabs can remain out of water up to four days. Crabs stranded on the beach during spawning bury themselves in the sand or fold themselves in half to conserve water until the tide rises again. Horseshoe crabs have no jaws or teeth. Instead, they have an impressive array of spiny mouth bristles at the base of five pairs of legs to maneuver food items such as razor clams, soft-shelled clams, and marine worms into their centrally located mouth. To chew its food, the crab must simulate walking movements.[215]

Another unique and intriguing feature of this ancient species is that it has copper-based blood, which turns blue (as opposed to red) when it encounters oxygen, in other words, they are "blue bloods." In the 1950s and 1960s, Dr. Frederik Bang, a Johns Hopkins researcher working at the Marine Biological Laboratory in Woods Hole, Massachusetts, found that when common marine bacteria were injected into the bloodstream of the horseshoe crab, massive clotting occurred. Later, with the collaboration of Dr. Jack Levin, the MBL team showed that the clotting was due to the presence of a gram-negative bacterial toxin called endotoxin. These investigators were able to localize the clotting phenomenon to the blood cells, called amebocytes, of the horseshoe crab, and, more importantly, to demonstrate the clotting reaction in a test tube. The cell-free reagent that resulted was named Limulus amebocyte lysate, or LAL. This reaction was then used to look for endotoxins in a variety of pharmaceutical and medical devices to ensure that they were not contaminated.[216, 217,218]

Why do horseshoe crabs have the ability to form clots around bacteria? Unlike mammals, the horseshoe crab lacks an immune system so it cannot develop antibodies to fight infection. However, the horseshoe crab does contain several compounds that will bind to and inactivate bacteria, fungi, and viruses. The components of LAL are part of this primitive "immune" system. For example, they not only bind and inactivate bacterial endotoxin, but the clot formed because of activation by endotoxin provides wound control by preventing bleeding and forming a physical barrier against additional bacterial entry and infection.[219]

Fast forward several years later and humans decided this blood was a commodity that needed to be harvested and used for a variety of medical needs. Today, LAL has become the worldwide standard screening test for bacterial contamination. Every drug certified by the FDA must be tested using LAL, as do surgical implants such as pacemakers and prosthetic devices.

Horseshoe crabs are collected using dredges or clam rakes, packed into trucks (sometimes refrigerated), driven to labs to have 1/3 of their blood removed, then returned to the ocean. Some studies estimate 10 to 15 percent of animals do not survive the bleeding procedure, which accounts for the mortality of 20,000 to 37,500 horseshoe crabs per year. Others have said those numbers may be double with up to 30% of horseshoe crabs bled dying.[220] Blood volume returns to normal in about a week, though blood cell count can take two to three months to fully rebound. A single horseshoe crab can be worth $2,500 over its lifetime for periodic blood extractions and a quart of extracted LAL was estimated in an article in 2011 in Wired magazine to be worth $15,000.

The worldwide market for LAL is currently estimated to be approximately $50 million per year.[221] Because of the lucrative market in their blood and the abuse of their collection as bait for conch and lobster fisheries, it's illegal in some states to harvest or even possess a live one without a proven scientific purpose.

Horseshoe crabs were the inspiration for the vocal effects provided by voice master Peter Cullen in the 1987 (and subsequent) "Predator" horror /science fiction movies. In an interview, Cullen says, "as I watched the Predator take off his helmet, I remembered the sounds of an upside-down horseshoe crab bubbling in the sun. The sounds of the clicking bursting bubbles came to me. The horrible underside of the dying crab and the face of the Predator just intertwined."[222] Rumor has it that H. R. Giger, the designer of the "Alien" xenomorph drew the "facehugger" based on the horseshoe crab anatomy, but that is a coincidence. The actual design was based on a combination of two hands and, well, if you must know, human "naughty bits."[223]

And this correlation brings us back to why they are so abundant along our beaches in May and June. Adults spend the winter in deep bay waters and off-shore areas. Horseshoe crab spawning season varies according to latitude, but it generally peaks in May and June, with peak spawning occurring on evening high tides during the full and new moons (the higher-than-normal "spring" tides).

The adults seek beaches that are at least partially protected from surf, within bays and coves. When the Limuli head for shore, the males patrol along the foot of the beach, awaiting the females. The female horseshoes give off chemical attractants called pheromones, which the males can detect. Although there may be other means of identification, these attractants, the directional movement, and the number of males involved (often several times the number of females) reduce the chance of a female reaching the beach without a boyfriend or two. Males, who are about 30% smaller than females, use a specially developed "boxer's glove" shaped appendage to clasp onto the back of the female. Female front appendages look a bit like feather dusters. Sometimes, several males will attempt to attach to a female and will form clusters with satellite males jostling for position. Males externally fertilize the eggs as the female deposits them. During our surveys, we find that as we walk around in the water with waders, our feet will become irresistible, female-like objects to any circling males which quickly demonstrate that sight is not their most highly developed sense.

By the beginning of the spawning season, each female will have developed about 80,000 eggs (in the Delaware region, in Massachusetts approximately 15,000-30,000 eggs depending on size), which are located in dense masses near the front of her shell. She will return to the beach on successive tides, laying 4-5 clutches of eggs with each tide.

Each cluster contains about 4,000 eggs and a female will lay about 20 egg clusters each year. Newly laid horseshoe crab eggs are opaque, pastel-green in color, and about 1.5 mm (1/16 inch) in diameter. It takes two weeks for the horseshoe crab to progress from egg to larvae to hatchling. If an egg is exposed to air for long it will dry out, but it will form an important source of food to migrating shorebirds.[224]

Juvenile horseshoe crabs generally spend their first and second summer on the intertidal flats feeding before the daytime low tide and burrowing in the sand for the rest of the day. As they grow, young crabs move into deeper water. Horseshoe crab eggs and larvae are a seasonal food item of invertebrates and fish. Striped bass and white perch eat horseshoe crab eggs. In addition, American eel, killifish, silver perch, weakfish, kingfish, silversides, summer flounder and winter flounder, most crab species and several gastropods including whelks eat eggs and larvae. Many sea turtles eat them; abundant stocks of adult horseshoe crabs may be an important component of ensuring the long-term survival of loggerhead sea turtles in the Delaware and Chesapeake Bays.[225]

Their eggs are a critical source of food for shorebirds, especially birds like the red knot. More than half of the total flyway population of red knots, ruddy turnstones and semipalmated sandpipers depend on Delaware Bay's horseshoe crab eggs as a rich food supply."[226] A reduction in horseshoe crab population cascades through the populations of the birds and other creatures who depend upon their eggs for food. A precipitous drop in red knot populations was the first clue we were impacting horseshoe crab populations.

Some folks aren't aware that horseshoe crabs molt (spiders do too). When you're walking in an area along the shore like a bordering marsh and you find all these little shells of horseshoe crabs, you're finding the leftovers after the crab wiggles out of its too-small shell in order to become larger. Before molting, a new shell begins to form. When this new shell is ready, the horseshoe crab absorbs water through its gills, making itself bigger. The old, hard shell cannot expand and splits in the front where the top and bottom join. The horseshoe crab crawls out the front, leaving the old shell behind. It takes about 24 hours for the new soft shell to harden. With each molt, the horseshoe crab increases in size by an estimated 25-30%. By the end of its first year, the crab will have molted on average at least six times, but will still be very small, clocking in at about one-half inch in diameter.

Horseshoe crabs initially molt an average of three or four times a year. Sub-adults (horseshoe crabs that are five to seven years old) appear to molt annually, usually in July or August. Males are sexually mature at their sixteenth molt, which is usually their eighth or ninth year. During their final molt, they develop specialized clasping claws for holding the female during reproduction. Females need at least 17 molts, or one more than the males, so they mature in their tenth year or even later and are, on the average, 30% larger than the males.[227] A small percentage of horseshoe crabs continue to molt after reaching sexual maturity. Scientists are not sure how long a horseshoe crab can live, but conservative estimates are at least 20 years.

Horseshoe crabs carry a variety of hitch-hikers during their journey, kind of like the world's oldest Country Squire station wagon. The typical 5-10 year old horseshoe crab has slipper shells and barnacles and algae growing on the top of the shell. I often told my students that for many of the mollusks and snails, the horseshoe crab is a convenient movable platform that helps keep them safe and moves them closer to food. Tiny crabs may hide along the inside "lip" of the shell and the Limulus leech (*Bdelloura*) is a flatworm that is found around the book gills and leg joints of crabs, especially on older females that have not shed for a long time.[228]

In 1990, the first organized survey of spawning horseshoe crabs in the Delaware Bay began. Now, every May and June during the full and new moon evening high tides, volunteers donate their time to count crabs on key beaches up and down the east coast. Each of the four peak spawning tides are bracketed with a count two days before and two days after, bringing the total number of survey nights to twelve.

Why are we concerned about their population? Besides the fact that it is frankly kind of rude to decimate an animal who has peaceably and successfully been existing for hundreds of millennia, the horseshoe crab eggs and larvae provide food for a huge variety of animals. Everything from migrating shorebirds to sea turtles, fish, crabs, and whelks dine on a horseshoe crab egg and larvae diet.

The huge drop in population along the east coast has been attributed to a variety of factors, including habitat alteration (bulkheads, groins, and associated development can remove naturally occurring beaches), overuse for medical purposes, harvesting for chitin, and overfishing for use as bait in the conch and eel fisheries and for fertilizer and animal feed. Three million individuals were killed in 1999 due to fishing mortality, i.e., using horseshoe crabs as bait. Bait bags and alternative bait gear is now used to wean fishermen off the previously abundant horseshoe crabs.[229]

Anecdotally, Nantucket Harbor and the surrounding water supported several thousand horseshoe crabs over the past 50 years. Those numbers have declined noticeably in recent years. Obtaining baseline data 20 years ago would have been ideal, getting it now is better than nothing.

People have harvested horseshoe crabs for centuries. Prior to the European colonization of North America, native tribes used the telson as spear tips and used the shell as

Photo from the author's archives.

containers. These small and localized harvests had little impact on horseshoe crab populations. In the late 1800s and early 1900s, up to 4 million horseshoe crabs were harvested annually and used as fertilizer or animal food. They were known as junk or trash fish, which is wrong on both counts. Currently, crabs are harvested for bait in conch and American eel fisheries on the Atlantic Coast. Horseshoe crabs suffered a substantial increase in harvest in the 1990s that spurred the need for management on a coast-wide scale. In 1998, the Atlantic States Marine Fisheries

Commission, representing 15 states from Maine to Florida, developed a horseshoe crab management plan. The ASMFC plan and its subsequent addenda established mandatory state-by state harvest quotas and created the 1,500-square-mile Carl N. Shuster Jr. Horseshoe Crab Sanctuary off the mouth of Delaware Bay. A combination of management efforts, research into alternatives for bait for conch fishermen and harvest quotas have started to very slowly turn the tide for these creatures. Their recovery will take some time because it takes a long time for each horseshoe crab to start breeding.[230]

Not much is known about where horseshoe crabs go (other than into deeper waters) when they are not entertaining us in the tidal waters. The horseshoe crab's main strategy to avoid predators is to be most active at night, feeding and spawning under the cover of darkness. In fact, during spawning season, you will find 100 times more crabs on shore at night than during the day. During high tide when large aquatic predators are swimming nearby, the juvenile horseshoe crabs bury themselves in the sand for protection. At low tide, young horseshoe crabs emerge from the sediment, but now they must be cautious of predators on the shore. If the crabs are turned upside down, they will use their telson to flip over, this movement always reminds me of a slow-moving windshield wiper. The need for them to keep their telson intact is the primary reason why you should never pick up a horseshoe crab by its tail. Instead lift it by the shell and flip it back over, avoiding the telson. Let's let them live to be around a few more millennia.

Strange & Unusual (Sea) Creatures
(August 20-26, 2009)

A life as the Director of a marine field station has many perks, not the least of which is the ability to encounter weird creatures the public does not normally see. One of my all-time favorite phantasmagorical creatures is a translucent, gelatinous stick that waves to you from the ridges of the mooring ropes it is most often found on—namely the skeleton shrimp. We would find them encrusting the lines and traps for oyster aquaculture cages in the Head of the harbor. They were also commonly found hanging around the Brant Point Marine Department shellfish grow-out facility. I had never heard of them before I laid eyes on them. These little guys are a bit creepy, but once you get used to them, they are really fascinating. Kids love them! They are clear and stick-like with a shape than can only be described as alien and bizarre.

Skeleton shrimp (caprellids) are amphipod crustaceans with very slender cylindrical bodies. Some people call them phantom shrimp. They look a bit like praying mantises or walking sticks and are often found clinging to sponges,

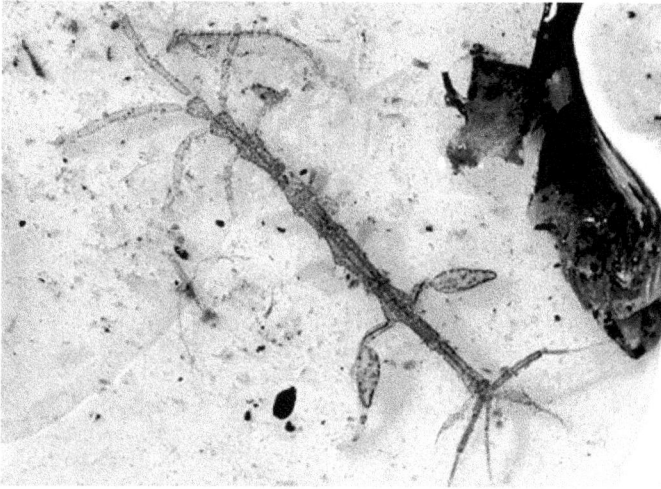

Skeleton Shrimp; photo by author.

hydroids, algae, and other aquatic organisms with appendages called pereopods.[231] The praying mantis resemblance is reinforced by their feeding technique as they face into the current with their clawed legs outstretched so they can capture drifting plankton. They can be very predatory, remaining motionless, looking like the eelgrass they are grasping, until an unsuspecting tiny invertebrate floats near them.[232] Some of them have developed their antennae to capture food. Like other amphipods, skeleton shrimp have two pairs of antennae, but the legs behind the first pair are greatly reduced in number. The first pair can be seen just under the head, the second pair carries the large grasping claws, and those at the hind end are used for holding on to the substrate. These animals can move by grasping alternately with the front and hind legs, like an

184

inchworm.[233] They can also swim by rapidly bending and straightening their bodies; one article described their swimming style as "thrashing" which sounds precisely like my swimming "technique." The female of some species kills the male after mating, and then carries her eggs in a brood pouch on the middle part of her body. The females typically undergo live birth.

These little guys are frequently found on ropes and other netting and are considered a biofouling organism in some anti-skeleton shrimp circles. Any fixed object—wharf piling, boat bottom, or dock—placed in the ocean rapidly becomes the site of an unusual assortment of animals known as the "fouling community."[234] First, barnacles settle out, then hydroids, sponges, bryozoans, and other sessile forms. These in turn provide a vast habitat for a multitude of small organisms, including our bizarre skeleton shrimp, tube building amphipods, isopods, and small crabs. Algae and tunicates appear and spread in orange, green, tan, and reddish mats.[235]

In this world flatworms, nudibranchs, and errant polychaetes crawl over the hydroids. Sometimes small nemerteans and nematodes appear, or tiny pink anemones. These fouling animals are a rich source of food and attract large numbers of fish. Under the dissecting microscope this collection provides a large diversity of invertebrate forms and demonstrates the concept of the artificial reef. Every year, students and interns at the UMass Boston Nantucket Field Station put out a variety of settling plates of different sizes and materials to see what type of organisms are floating around, looking for a spot to settle. They are always on the lookout for exotic invaders who may have entered Nantucket or Madaket harbors latched onto the hulls of visitors or floating around in ballast tanks.[236]

There is a notorious species that has been invading harbors and has traveled far and wide. It is the Japanese skeleton shrimp (*Caprella mutica*) and it has spread from its Pacific origin in the waters off northeastern Japan and Russia to twenty-nine non-native locations around the globe, spanning both hemispheres. *Caprella mutica* has been found in the British Isles, Ireland, Norway, Germany, Belgium, the Netherlands, North America, and New Zealand.[237] Whether our local skeleton shrimp are predominantly this very successful invader is unknown and investigating the level of incursion would be a great undergraduate project.

Small Monsters in the Water – Chimeras
(August 16, 2012)

A chimera (from the Greek word for she-goat) is a fire breathing female creature from Greek mythology made of three different animals; a lion, a serpent and a goat. The chimera was the "daughter" of Typhon and Echidna and the "sister" of Cerberus and the Hydra. The chimera was destroyed by Bellerophon atop his steed, Pegasus, a better-known chimera. In literature, the term chimera is used to describe any mythical or fictional animal composed of parts of other animals like the griffin or Minotaur.[238]

In genetics a chimera is a single organism (usually an animal but can be a plant) composed of two or more different populations of genetically different cells that come from two different zygotes involved in sexual reproduction. In other words, they form when two embryos fuse together in the womb or seed and the result is a mix of tissues. This occurs in humans and is more common in in-vitro fertilization. It can even result in a person having two different blood types or both types of sexual organs and this condition is not as rare as originally beloved.[239] Almost all marmosets are chimeras, specifically germline chimeras, as a result of being born as fraternal twins and sharing DNA with their sibling.[240]

Anglerfish exhibit probably the strangest case of "parasitic chimerism" as a "natural" part of their life stages. This is one of those "truth is stranger than fiction" stories. Male Ceratioid Anglerfish (deep sea fish also called Sea Devils) start searching for a female anglerfish from the moment they are born. They find the females using super developed olfactory glands and latch on to the much larger females. Then these little males (sometimes can be more than one) release an enzyme that begins to digest the skin and scales of both their mouthparts and the female body down to the blood vessel level until they fuse into a hermaphroditic individual.[241] This is crazy isn't it? This process is a reminds me of the hermaphroditic slipper shells bivalves. There is a method to this madness. Once they are fused to the female; the males develop into sexual maturity becoming essentially one giant set of testicles that supplies a constant amount of sperm that fertilizes the Borg-like females' so that more offspring can be produced. Why this is not a Ridley Scott movie, I don't know.

Photo by the author

Today's subject reminds me of the mythical not biological versions of chimeras in which two or more creatures are fused together. The mantis shrimp is a large invertebrate crustacean with praying mantis-like front claws. They are neither shrimp nor mantids. One summer, one of our youngest visitors (who was almost three) found a HUGE mantis shrimp in the intertidal area up on the beach (see the picture with a ruler juxtaposed for size). This gave the junior rangers, interns and I a chance to do some detective work. From "A Practical Guide to the Marine Animals of Northeastern North America" by Leland W. Pollock, we learned as a team that the mantis shrimp's Latin name is *Squilla empusa* (which sounds a bit like a Spongebob Squarepants character). They are in the Superorder Hoplocarida, Order Stomatopoda, Family Squillidae, and they are often known by their order name Stomatopods (which is from Greek "stoma" mouth + "-podos" for foot) These relatively large invertebrate critters (ours was between 8.5 and 10 inches long depending on where you measured it) are dorso-ventrally flattened (like a cross between a crawfish and a lobster or a creature who wants to win at limbo-ing). They are yellowish green with pink tinges and bright green eyes.[242]

S. empusa are nocturnal and live in U shaped burrows emerging to feed on crustaceans and fish. They like to dig their burrows in deep mud, and they may have multiple openings to the surface spread two or three feet apart. They are aggressive and are relatively common from the Cape Cod region down to Gulf of Mexico and off the coast of Brazil and in the Mediterranean Sea.[243] Most other species of mantis shrimp are tropical or subtropical. Adult *S. empusa*

189

can grow to a length of 8-10 inches or 30 cm. long, making our specimen full grown and most likely past the point of needing to molt. Their long, flattened bodies are segmented and tend to be translucent to light. Their body is divided into 2 main parts, a cephalothorax (head fused with the thorax), and an abdomen that ends with a telson. The abdomen is broad and fully developed, fanning out towards the end. It is divided into 5 clear segments that are outlined by a dark greenish bluish color or sometimes yellow (these colors help them recognize other mantis shrimp). Attached to the abdomen's middle line are several pairs of pleopods or swimmerets used for swimming, which also have special filaments and gills for respiration. The last pair of appendages is the uropods found at the sides of the tail or telson. Their telson is covered by 6 sharp spikes and is highly flexible and is normally used to fend off enemies and other mantis shrimp.[244]

I bet you didn't know that mantis shrimp have the fastest punch of any animal in the world. According to the web site "Not exactly Rocket Science," "Mantis shrimps are mere inches long but can throw the fastest punch of any animal. They strike with the force of a rifle bullet and can shatter aquarium glass and crab shells alike. Now with the aid of super-speed cameras, we are beginning to truly appreciate how powerful this animal is. Their secret weapons are a pair of hinged arms folded away under their head, which they can unfurl at incredible speeds. The "punch" can be delivered in two different manners depending on the species of mantis shrimp. The "spearer" species have arms ending in a fiendish barbed spike that they use to impale soft-bodied prey like fish. But the larger "smasher" species have arms ending in heavy clubs and they use them to deliver blows with the same force as a rifle bullet."[245] A four inch long (remembering our specimen is 10 inches long) mantis shrimp managed to break a ¼ inch glass aquarium tank and escape. Ironically, this shrimp was named Tyson. The researchers had a heck of a time finding a fast motion video or still camera that could capture the creature's movement. Sheila Patek, a researcher at USC Berkeley, was able to finally record these species with the help of a BBC crew and their special fast motion video camera used for the television series *Animal Camera*. As an aside, Dr. Patek also discovered that the fastest limb movement of any animal belongs to the trapjaw ant whose mandibles close with an almost unbelievable maximum speed of 140 mph. The mantis shrimp came in second at 50 miles per hour. And this lightning-like strike is through water which is much denser than air. The mantis shrimp does this using its hinged arms, which, when pulled back, are latched with a ratchet-like device that compresses a cuplike (or Pringle chip shaped) spring that stores energy. These are unleashed in a punch that is much stronger and faster than simply muscles extending outward. In the field of biomimicry, this naturally developed spring lock action could be used to develop superior machines.

Patek's cameras revealed that each of the smasher's strikes produced small flashes of light that occur because the club moves so quickly that it lowers the pressure of the water in front of it, causing it to boil. The water then releases bubbles as the water pressure normalizes within a blink of an eye, causing cavitation or turbulence that further harms their truly hapless victim. Even the thick exoskeleton of a crab is no match for this force and fish are speared expertly as if the mantis shrimp wielded a spear gun. For the *S. empusa*, the sharp claw is assisted by 6 spines found at the last joint which gives the claw extra slicing abilities that have been known to cut human flesh which gives these creatures some of their many nicknames such as "thumbsplitters".[246]

The "amazingness" (is that a word?) does not end there. Mantis shrimp have the most evolved eyes in the animal world. Some stomatopods can see polarized light and even parts of the UV spectrum. They have 16 different types of photoreceptors, 12 of which are devoted to color (the human eye has 3) and 4 for UV light, allowing them to perceive a total of 100,000 different colors. Some scientists in Australia have documented that the species of mantis shrimp with the most highly evolved visual systems can even detect types of cancer![247] The *S. Empusa* can see polarized light but not colors which is an adaptation to the murkier sediments it inhabits (Cronin et al. 1994).[248] Last but not least, *S. empusa* is extremely intelligent, very belligerent, a solitary animal and one of the spearing type Stomatopods. They can solve problems and figure out simple puzzles despite having no real brain. They also recognize others of their species and rivals up to a month later assisted by the colors lining the exoskeleton mentioned above. In fact, researchers have found that some mantis shrimp (mainly the colorful cousins) have "mushroom bodies" in their brains similar to insects that allow them to make these neural connections.[249] And they are extremely aggressive, more so than other similar sized crustaceans like lobsters. Their fast reflexes, which include jumping back from predators and flipping around in a somersault to run away, to punching or spearing their prey, result from a series of ganglia running along their long bodes which allow them to respond extremely quickly to a variety of stimuli. The more one learns about these creatures, the more interesting they become, although many aquarists detest these wily, hungry, aquarium-escaping creatures.[250]

Mantis shrimp are eaten in many countries including the Philippines, Japan, Vietnam, in the Mediterranean, China and Hawaii. I think I need to write a script for a monster movie based on the mantis shrimp, maybe a cross between "When Tyson met Predator."

Section 3: flotsam and jetsam - the washashores

Floating Wonders & Blue Bottle Washashore
(Aug 28-Sept 3, 2008 and June 13, 2013)

As the summer temperatures warm up our harbor and offshore waters, floating translucent creatures begin to crowd our shores and worry swimmers. Unfortunately, as ocean waters warm world-wide and we continue to overfish or endanger their natural predators, the occurrence and number of jellyfish are on the rise. Fortunately, our colder waters keep many of the extremely painful species relatively far from our beaches, although a few interlopers sneak in.

Jellyfish are the most primitive of the multicellular organisms. Although they can "swim" or move under their own power in a form of locomotion, they are more influenced by tides and currents and almost planktonic in their motions.

They are famous for their beautiful transcendent appearance in aquariums and infamous for their stinging capsules contained within cells called cnidocytes located along the tentacles. These cells act as a trigger with a toxin bullet composed of a protein-based venom inside the cnidocyte. A cnidocyte fires a structure that contains the toxin, from a sub-cellular organelle called a cnidocyst (also known as a "cnida," Latin for "nettle")[251] or nematocyst. This is responsible for the stings delivered by jellyfish. This acts as a one-shot harpoon that literally embeds itself within tissue injecting the toxin.[252] Scientists are discovering that these cells evolved when cnidarians, specifically a species of sea anemones, repurposed ancestral neural cells to make these novel stinging cells. This research leads to many new and interesting ways to track how cells evolve and whether they come from stem cells or other previously assigned cell tasks.[253]

As a chemist, and a chemical oceanographer, I think in terms of chemistry more than biology. Knowing that this chemical is protein based is helpful for counteracting it. When I worked and lived in Galveston, Texas (which was basically jellyfish heaven), we always carried Adolph's meat tenderizer or some type of papain-based spice to help break down the proteins in the stinging cells. But nowadays, the old standby treatments like ammonia, vinegar, meat tenderizer, and urine are not considered nearly as effective as hot water or a heat pack (not that anyone has that on them) or lidocaine or other topical anesthetics. The Red Cross (and I) advise using salt water first to rinse the area as it does not cause triggering of the nematocysts. You should also apply a poultice of baking soda mixed with salt water or talc plus saltwater to extract the stinging cells.[254,255]

Thankfully, many of the smaller species cannot pierce human skin or produce enough of a sting to affect us. They concentrate their stunning power on smaller fish, plankton, and sometimes each other. In a way, jellyfish can "hunt," feeling their way along the ocean currents, using rudimentary light and pressure sensors to detect victims, and silently dispatching their dinners. Typically, only when they become a nuisance, or a danger do we become aware of them.

Jellyfish are invertebrates (no backbone) animals in the phylum Cnidaria in the family Medusae. There are four major groups of cnidarians: the sessile (immobile) Anthozoa, which includes true corals, anemones, and sea pens; Cubozoa, the amazing box jellies with complex eyes and potent toxins; Hydrozoa, the most diverse group with siphonophores, hydroids, fire corals, and many medusae; and swimming Scyphozoa, the true jellyfish.[256]

Many cnidarians are gelatinous and all of them are radially symmetrical. Radial symmetry aligns all the body parts around a main axis in a cylinder. One significant feature of radial symmetry is that it allows an animal to confront their environment in numerous directions. A radially-symmetrical animal has no front or back end. This body form is most common in sessile and drifting species who don't need to go anywhere in a straight line.[257] Jellyfish are theorized to have appeared in the oceans about 650 million years ago, before the dinosaurs. Known for being squishy, not bony, few jellyfish have been preserved as fossils. The oldest known one was found as an impression in a rock dating back to the Middle Cambrian 505 million years.[258] Jellyfish are made up of 95% water; they have no bones or cartilage, no heart or blood, and no brain! Now you know what to call your little brother or sister.

In a rare occurrence in the summer of 2006 (right before the 4th of July), hundreds of Portuguese man-of-war (also spelled man o'war) jellyfish (*Physalia physalia*) showed up on southern Nantucket beaches, which prompted the temporary closing of many popular swimming beaches.[259] Although a few individuals can float into our waters, such a large influx can usually be contributed to a shift in the Gulf Stream. Man-of-wars have a multicolored balloon-like float with tentacles that can hang down to 50 feet. The easiest way to explain what a dead one looks like is to compare it to those the balloons (highly toxic) created by blowing a purplish compound through a thin plastic straw popular in the last century. Portuguese man-of-wars are not jellyfish, but instead are a colonial animal composed of polyps and medusoid individuals called a siphonophore (Order Siphonophora) and are a member of the hydrozoa, or hydroid class. The medusa-form body consists of a translucent, jellylike, gas-filled float, which may be 3 to 12 inches long. Polyps beneath the float trail tentacles up to 165 feet long. Nematocysts on some polyps paralyze fish and other prey. Other polyps then attach to, spread over, and digest the victim.[260] The Portuguese Man O' War (named caravela-portuguesa in Portuguese) is named for its air bladder, which looks similar to the sails of the Portuguese fighting ship (Man of war) caravela redonda of the 14th and 15th centuries.[261] The Portuguese man o' war, is an excellent example of a members of the *neuston*. The neuston is the floating community of ocean organisms that live at the interface between water and air.[262]

Man o' wars can deliver a very serious sting and it is not uncommon to have the tentacles wrap around your entire body. Some kinds of jellyfish and *P. physalia* can also still sting while washed up on shore because the nematocyst "trigger" has not fired yet; so be sure to avoid them (dead or alive).

Moon jellies (*Aurelia aurita*) are the most common member of the jellyfish family seen around Nantucket and on North Atlantic shores. Moon jellies are translucent with 4 pink horseshoe-shaped gonads on the dorsal (top) of the creature and very short tentacles. Moon jellies are relatively small and do not sting humans. The threadlike tentacles around the edge of the bell can sting, and may occasionally catch small swimming animals for food, but their nematocytes are not powerful enough to pierce our thick skin. They feed mostly by trapping microscopic plankton in a film of mucus which flows over the surface of the bell and is picked off as it reaches the edges by the thick mouth tentacles underneath. They swim by pulsing their bell, pushing themselves slowly through the water.[263]

Occasionally our surrounding waters may contain lion's mane (*Cyanea capillata*) and stinging Atlantic sea nettles (*Chrysaora quinquecirrha*), both of which are true jellyfish. The lion's mane jellyfish is the largest known jellyfish species and prefers colder waters. You may remember this jellyfish from the Sir Arthur Conan Doyle Sherlock Holmes story, "The Adventure of the Lion's Mane" published in "The Case-Book of Sherlock Holmes." The largest recorded specimen had a bell (body) diameter of 2.3 m (7 feet 6 inches) and tentacles 36.5 m (120 feet) long- longer than the longest blue whale recorded! It was found washed up on the shore of Massachusetts Bay in 1870.[264]

Locally, nowadays, the juvenile lion's manes show up in late spring and are too small to really bother swimmers; they grow larger and less common throughout the summer and reach a typical bell size of 12 inches. The lion's mane starts as a pinkish juvenile and their color deepens to a purple or red as they age, although some smaller adult individuals are light orange or brown. The bell is divided into eight lobes, giving it the appearance of an eight-pointed star. Each lobe contains about 70 to 150 tentacles, arranged in four rows. A tangled arrangement of colorful arms emanates from the center of the bell, much shorter than the silvery, thin tentacles which emanate from the bell's subumbrella. These jellyfish are understandably named for their showy, trailing tentacles reminiscent of a lion's mane.[265] The lion's mane sting is moderately painful and sometimes feels like a burn. It is important to note that while jellyfish stings can be excruciating, they are rarely serious health threats unless one is allergic to them.

The Atlantic sea nettle looks very much like lion's mane. This jellyfish is saucer-shaped with brown or red pigments, usually 6-8 inches in diameter. Sometimes they are semitransparent or will have reddish stripes. Four oral arms and long marginal tentacles hang from the bell and can extend several feet.[266] Considered moderate to severe, symptoms from sea nettle stings are like those of the lion's mane.

Sea nettles spawn in mid to late summer and they usually die after spawning. First the males release sperm into the water, then the females' eggs are fertilized as they swim and pump water through their body. Not the world's most efficient process, but whatever floats your boat. After fertilization, the eggs develop into tiny, free-swimming larvae called planulae, which the female releases into the water. The larvae float with the currents for a few days, then settle and attach to a firm surface. The larvae blossom into anemone-like polyps that bud and grow over the winter. By spring, the polyps develop tiny, floating medusae that are layered on top of one another. The medusae are eventually released into the water. The freely floating medusae (called ephyra) eventually grow tentacles and mature into adults.[267] One of the key steps is the attachment of the larvae and the increasing number of hard surfaces like bulkheads and docks has been blamed for an increase in jellyfish populations in some embayments.[268] Other factors like excess nutrients, warming waters, and a reduction in their natural predators likely play a large part in ballooning jellyfish populations.

My favorite jellyfish-like creature is the comb jelly, which is not a jellyfish at all, but instead a member of the Ctenophora (pronounced "teen o four a") phylum. Comb jellies differ from jellyfish in that they have 8 rows of comb-like cilia ("tiny hairs") which are paddle-like structures that beat in the water to aid in orientation and locomotion. Comb jellies also have biradial symmetry. Comb jellies do not sting! These little guys are delightful creatures also called "sea snot," the reason for which will become abundantly clear if you ever hold one in your hand.

They are quite beautiful in the daytime with the cilia pulsing and scattering light producing a rainbow type effect, and they are even more beautiful at night because they produce a green light using chemical reactions in a process called bioluminescence that causes these creatures to flash on and off when disturbed. You can see this light show on dark nights in the harbor especially near the Head of the Harbor and off Steps Beach. Often in July, our marine class plankton nets fill up with comb jellies in just a few seconds.

In the summer of 2008, our most common comb jelly, the Leidy's comb jelly (*Mnemiopsis leidyi*, also known as the sea walnut), arrived even sooner than usual. Unfortunately, a large influx of Leidy's comb jellies has been causing a significant food chain collapse in the Black Sea for close to a decade as they out-compete fish such as anchovies for food items such as zooplankton.[269] The comb jellies and other jellyfish have an advantage over the visual hunters, like fish, as the waters in our seas and harbors become eutrophic (plankton-laden) and light penetration decreases; remember that jellyfish "hunt" tactically, so the lack of light is less of a problem for them. In the warm waters of the Black Sea, these ravenous carnivores eat the zooplankton before the fish eggs can hatch and the larval fish species can begin to eat. These comb jellies came over as an invasive non-native organism in the early 1980's in ship ballast water.[270] Currently, the only option for fighting these is the introduction of a different comb jelly species, *Beroe ovata*, which is a predator of the Leidy's comb jelly. Some of the

Black Sea fisheries are beginning to rebound as the Leidy's comb jelly population decreases.[271]

Jellyfish are a favorite menu item for many sea turtles, which is why the abundance of balloons and plastic bags floating in the ocean have found there (sometimes lethal) way into a turtle's stomach. One of the largest sea turtle species, the leatherback, consumes jellyfish almost exclusively. Ocean sunfish (*Mola mola*), some tuna and shark species, and ocean birds such as fulmars and phalaropes also eat jellyfish and are immune to their toxins. Jellyfish are also on our dinner plate in popular Japanese and Chinese dishes. I've had it once and survived (it was quite good, a pleasant, jiggly, cold pickled salad).

The Kraken's Younger Brother & Abundance of Squid
(June 24-30, 2010 and June 19, 2014)

I was visiting an island seafood purveyor in early June 2010 when the topic of squid came up. I am a big fan of squid and other less traditional seafood and was wondering if there was viable local market for them and how sustainable the fishery might be. I knew people locally were into jigging for squid on the docks, but I was curious as to whether they were sold frequently. The owner told me very few commercial fishermen fish for squid and what catch there is depends on the use of small gauge nets for winter flounder and the ongoing fishing regulations. Mid-June, fishermen switch to nets with larger holes that the squid can slip through. Although this practice makes it difficult to stock and sell this seafood in stores, it does help protect the breeding stock. We'll talk more about the fishery below.

Squid are marine cephalopods in the superorder Decapodiformes, which includes all species with ten limbs; the name derives from the Greek word meaning ten feet. The ten limbs are divided into eight short arms and two long tentacles. It is hypothesized that the ancestral coleoid had five identical pairs of limbs, and that one branch of descendants evolved a modified arm pair IV to become the Decapodiformes, while another branch of descendants evolved and then eventually lost its arm pair II, becoming the Octopodiformes (octopi).[272,273]

Like all other cephalopods, squid have a distinct head, bilateral symmetry, a mantle, and arms. Squid, like cuttlefish, have eight arms arranged in pairs and two longer tentacles (for a total of ten limbs). Tentacles are longer than their arms and they usually only have suckers at their tips. Squids are mollusks of the phylum Mollusca. Their molluscan relatives are snails, clams, and other bivalves.

Squid are members of the class Cephalopoda. Other members include octopus, cuttlefish, and the ancient nautilus. The word Cephalopoda means literally, "head-foot bearing" and refers to the way the cephalopod arm appear to be attached directly to the head. The shell of the mollusk has evolved to become a cartilaginous plate shaped like a quill pen and buried under the mantle. The mantle, the chief swimming organ of the animal, is modified into lengthwise fins along the posterior end of the body and projects forward like a collar around the head. As the mantle relaxes and contracts, the squid swims forward, upward, and downward. Water is expelled in jets from the muscular funnel located just below the head, propelling the squid backward in abrupt jet-like motions. The combination of chitin and mantle makes the squid fast and able to withstand great depths and pressure. The arms are used to steer while swimming and the tentacles are used to seize and immobilize prey, which are then cut into pieces by the animal's strong beaklike jaws.[274] The beak is made of chitin and is indigestible which explains why squid beaks are often found in the stomachs of predators from whales to seals.[275]

The squid you most commonly see off Nantucket is one of the many species in the genus *Loligo* which is one of the most common and widely distributed squid species. Squid are most abundant in Nantucket Sound in May and June. On June 1, in 2014, squid eggs were reported at Great Point, Smith Point, and other beaches around Nantucket. Females lay hundreds of eggs in capsules that they attempt to attach to the sea floor along with their sister squid eggs; but these often come loose and wash ashore.

The genus was first described by Jean Baptiste Lamarck in 1798. However, the name had been used earlier than Lamarck and was most likely used by Pliny the Elder, also known as Gaius Plinius Secundus (circa 23-79), the Roman officer and encyclopedist in book 9 of his 37-volume opus called Naturalis Historia.[276]

Squid were long prized by scientists for their huge axons. More than 100 times the size of a human axon, the squid's axon is roughly a millimeter in diameter and the largest nerve fiber on Earth. The enormous size allows scientists to study how signals are sent along the length of the axon to other nerve cells or muscles. Squid research today focuses on everything from evaluating the progression of Alzheimer's damage in the brain to investigating cures for various eye diseases.[277]

In 1963, the Nobel Prize in Physiology or Medicine was awarded jointly to Sir John Carew Eccles, Alan Lloyd Hodgkin and Andrew Fielding Huxley "for their discoveries concerning the ionic mechanisms involved in excitation and inhibition in the peripheral and central portions of the nerve cell membrane." In essence, these men used squid to investigate how human neurons worked; essentially doing research on an axon large enough to see with the naked eye. In a serious of groundbreaking experiments that could have come out of the novel "Frankenstein: or, The Modern Prometheus," they hooked up electrodes and circuitry to the axons to see how electricity, and therefore information, could be transmitted.[278]

The longfin inshore squid (originally *Loligo pealeii*) is a species of squid of the family Loliginidae. It has recently been renamed the *Doryteuthis (Amerigo) pealeii*, and this is the species we see most often around Nantucket. The longfin inshore squid is found in the North Atlantic, schooling in continental shelf and slope waters from Newfoundland to the Gulf of Venezuela. It is commercially exploited, especially in the range from Southern Georges Bank to Cape Hatteras. *D. pealei* can reach lengths of up to 24 inches.[279] A close relative, the boreal squid, *Illex illecebrosus*, also known as the short-finned squid, overlaps *Loligo* over some of its range. It is generally found farther north. The two squids can be distinguished by the relative lengths of the fins. [280]

Squid are somewhat migratory creatures that comes to Cape waters every spring to spawn. They live offshore in deeper cooler water during the winter. If you have ever seen squid reproduction, you would certainly have not forgotten it. In early spring, squid "dating" starts as the dominant males fight for the opportunity to copulate with the females. The female gathers the sperm from one or more partners in the off-shore mating ritual. When she arrives inshore, the female pulls a jelly-like matrix of eggs out of her body and holds it her arms and releases the stored sperm to fertilize hundreds of eggs. It has been documented that sneaker males may lay in wait for females to deposit their egg sacs and fertilize the eggs at the last second, beating out the older sperm the female collected earlier.[281]

Most scientists agree that *D. pealei* only live on average for one year (sometimes only 6-8 months), although this conclusion was only recently reached as a result of studies of the animal's statocyst, a calcium carbonate rock in the animal's balance organ that grows in rings that are laid down daily.[282] Individuals hatched in summer generally grow much faster than those hatched in winter. The species is sexually dimorphic, with most males growing faster and reaching larger sizes than females. The dorsal-mantle length of some males can reach 40 cm, although most squids commercially harvested are smaller than 30 cm long.

Scientists at Woods Hole have discovered that the males freak out and get very aggressive when they encounter eggs. Researchers first noticed this effect while studying male squid arriving at breeding grounds where females had already started laying egg capsules. Each egg capsule contains 150 to 200 eggs. A female squid will lay 20 to 30 capsules over a period of up to several weeks, during which time she will mate with multiple males. The scientists noticed that, when male Longfin Squid saw egg capsules laid on the seafloor, the squid would swim toward the eggs and wrap their arms around them. They would then become enraged and attack each other.[283]

Squid hatchling will eat the yolks from their egg sacs for 4-5 days, then progress to eating copepods and other zooplankton, eventually graduating to fish and shrimp as they get larger. Cannibalism has also been documented for squid.[284] Many creatures like gray seals, stripers, and common dolphins feed on squid. Large groups of squid aggregate both south and north of the island each summer; sometimes bringing squid loving marine mammals in their wake like the pygmy and dwarf sperm whales seen last summer.

Squid are intelligent animals, for example, groups of Humboldt squid hunt cooperatively, using active communication.[285] Their skin is covered in chromatophores, which enable the squid to change color to suit its surroundings, making it effectively camouflaged. Controlled by the nervous system, the camouflage can change in a split second and their shoaling behavior with rapid color changes is quite beautiful.[286,287]

Squid breathe through gills, and, like its cousin the octopus, will squirt a cloud of inky material from its ink sac when in danger. The squid is one of the most highly developed invertebrates, well adapted to its active, carnivorous, predatory life. All squid have a mouth with a radula which is a file-like organ that scrapes nutrients from food sources. Squid will eat small copepods and other zooplankton when they are first born, graduating to fish and shrimp as they get larger. Lots of things eat squid, they are a major part of the food web. Sharks, ceteceans, such as beaked whales, gray seals, stripers, and common dolphins all feed on squid. Large groups of squid aggregate both south and north of the island each summer; sometimes drawing squid-loving marine mammals.

The circulatory and nervous systems are highly developed and the eyes of the squid are extremely evolved and often studied. The eye of the squid is remarkably like that of humans—an example of convergent evolution, as there is no common ancestor. The images squid see through hard lens on each side of the head is focused by changing the position of the lens, as in a camera or telescope versus changing the shape of the lens, like humans. Giant squid, which measure 40 feet and longer, and their smaller cousin, the *D. pealei*, have large optic nerves.[288] Some deep-sea forms have luminescent organs. More squid fun facts: they have three hearts and limited hearing. Probably the most reported squid fact is one we can't go into a lot of detail here, but let's say, except for a type of barnacle, deep sea male squid (specifically greater hooked squid Onykia ingens) are the most well-endowed of all creatures.[289]

Of course, no article about squid can forgo a discussion of the mythological creature, the Kraken. The Kraken of legend is probably what we know today as the giant squid (*Architeuthis dux*). While a colossal octopus might also fit the description, the squid is thought to be much more aggressive and more likely to come to the surface.

Many of my college friends worked at a squid behavioral center and neurological facility in Galveston at the University of Texas Medical Branch and they would spend their evenings watching hours of tapes of squid and cuttlefish swimming around in big tanks (yes, they got paid for it, college students must make money somehow). Each night, some wayward squid would tire of the tank and escape to be found the next morning in another tank or sometimes manage to open a door handle and be halfway down the hallway!

In the United States tons of squid are used for fish bait, particularly by the cod fisheries in New England. Many squid species are attracted to lights; they are therefore fished using different light attraction methods. On several oceanographic cruises that I have been on, we use bright working lights at night that would attract huge schools of squid in the Gulf of Mexico. Fishermen have been using their propensity to come to lights for many years. Squid are also a part of the Cape and Islands recreational fishery.

Fishing for squid (jigging) from lighted spots of docks or off boats is a very popular activity on both sides of the Sound.

Reading through riveting copies of the Marine Fisheries Advisor over the past years, we can see that the Division of Marine Fisheries for Massachusetts had changed regulations for trawling for squid. The squid regulations have evolved over time, with changes in the season, limitations on the size of trawlers in some key squid spawning areas, controlling the amount of discarded squid through the process of trawling for other species, and monitoring the size of squid caught.

Illex and Longfin squid are regulated under National Oceanographic and Atmospheric Administration (NOAA) Fisheries Division and governed by the "Atlantic Mackerel, Squid and Butterfish Management Plan" by the Mid Atlantic Fisheries Management Council in federal waters. Currently the season is divided into three or four periods with set quotas for some seasons and no limits during other seasons, effectively allowing the fishery to be open with few limits on size or number caught. Squid populations are replaced twice a year so fishing pressure is normally not a problem for them if some squid are left in the population at any given time. Spacing the allowed harvest throughout the year ensures fishing pressure isn't concentrated too heavily at one time and allows the fishery to operate year-round. NOAA fisheries stock assessments show that longfin squid is not overfished. U.S. wild-caught longfin squid is a smart seafood choice because it is sustainably managed and responsibly harvested under U.S. regulations.[290]

The take of long fin squid is managed via a coast-wide quota that varies from 15-45 million pounds of which Massachusetts takes on average 10-15 percent of the total catch. The variances in the quota are reflected in the boom/bust years of the squid. In 2019, commercial landings totaled more than 27 million pounds, and were valued at approximately $43 million, according to the NOAA Fisheries commercial fishing landings database.[291]

My obsession with squid started with the idea of having them for dinner. You can eat every part of a squid except for the beak and pen. On our restaurant plates, squid is known as calamari, which is the plural form of the Italian word for squid, Calamaro. Also known as Kalamari, Kalamar (Greek/Turkish), Calmar (french), Galama or Calamares (Spanish), the name derives from the Latin word calamarium for "ink pot," after the inky fluid that squid secrete. "Calamarium" in turn derives from Greek "kalamos" meaning "reed," "tube" or "pen," referring to the quill "pen" inside of the squid.[292] People use the ink to color food like paella and pasta, but also for writing. This sustainable seafood sustains many large creatures in New England and keeps the world fed too.

Notes

Chapter 1: Coastal Processes and weather

"Living with the Sea"

(Oct 1, 2009)

https://yesterdaysisland.com/archives/science/21.php

1. Hubert Cancik and, Helmuth Schneider, English Edition by: Christine F. Salazar, Classical Tradition volumes edited by: Manfred Landfester, and English Edition by: Francis G. Gentry, eds. 'Figura Etymologica'. In Brill's New Pauly. Accessed August 17, 2022. doi:http://dx.doi.org/10.1163/1574-9347_bnp_e411490.

2. Gary Martin, "The meaning and origin of the expression: Time and tide wait for no man," *The Phrase Finder*. Accessed August 10, 2019, https://www.phrases.org.uk/meanings/time-and-tide-wait-for-no-one.html.

3. Kathryn Westcott, "Is King Canute Misunderstood?" BBC News. May 26, 2011. https://www.bbc.com/news/magazine-13524677

4. "NOAA Technical Report NOS CO-OPS 051 Elevated East Coast Sea Level Anomaly: June – July 2009" National Oceanographic and Atmospheric Administration, accessed August 10, 2019, https://tidesandcurrents.noaa.gov/publications/EastCoast SeaLevelAnomaly_2009.pdf.

5. National Oceanographic and Atmospheric Administration, public release, 2-SEP-2009. "NOAA report explains sea level anomaly this summer along the US Atlantic coast," Accessed August 11, 2019, https://www.eurekalert.org/pub_releases/2009-09/nh-nre090209.php.

6. NOAA National Centers for Environmental Information, "State of the Climate: National Climate Report for August 2009" published online September 2009. https://www.ncdc.noaa.gov/sotc/national/200908.

"Winter Weather"

(November 14, 2013)

https://yesterdaysisland.com/winter-weather/

7. Weather Underground Station, Nantucket Memorial Airport. accessed on August 11, 2019, https://www.wunderground.com/history/daily/KACK/dat e/2013-11-10?req_city=&req_state=&req_statename=&reqdb.zip=&r eqdb.magic=&reqdb.wmo= Data.

8. National Oceanic and Atmospheric Administration's National Weather Service. "NOAA Climate Data" Accessed August 12, 2019. http://www.nws.noaa.gov/climate/index.php?wfo=BOX.

9. A. John Arnfield, "Köppen climate classification," *Encyclopædia Britannica*, Last modified April 11, 2019. https://www.britannica.com/science/Koppen-climate-classification.

10. "Nantucket June Average Sea Temperature" World Sea Temperature 2022. Accessed August 17, 2022, http://www.seatemperature.org/north-america/united-states/nantucket-june.htm.

11. U.S. Environmental Protection Agency, "What Climate Change means for Massachusetts," August 2016. EPA publication 430-F-16-023. https://19january2017snapshot.epa.gov/sites/production/files/2016-09/documents/climate-change-ma.pdf.

12. U.S. Environmental Protection Agency. 2016. "Climate change indicators in the United States, 2016,"

Fourth edition. EPA 430-R-16-004. Accessed August 11, 2019, https://www.epa.gov/sites/production/files/2016-08/documents/climate_indicators_2016.pdf.

13. U.S. **Environmental** Protection Agency, "What Climate Change means......" August 2016.

"Fog Happens"

(2011 Author's archive)

14. American **Meteorological** Society, "Fog" *Glossary of Meteorology*. Accessed August 11, 2019, http://glossary.ametsoc.org/wiki/fog.

15. Ministry of Defense. Admiralty Manual of Navigation, Volume II. London: **Her** Majesty's Stationery Office. (Navy, 1973) p. 188.

16. Susan Higgins, "5 Foggiest Places in North America" Farmer's Almanac, Posted March 16, **2015** https://www.farmersalmanac.com/5-foggiest-places-north-america-20735 .

17. **Geoffrey** Migiro, "The Foggiest Places on Earth." World Atlas. Accessed August 25, 2019, https://www.worldatlas.com/articles/foggiest-places-on-earth.html

18. Sarah Oktay, "When the Going Gets Hot" *Yesterday's Island* Volume 40, Issue 13. (July 29, **2010**). Accessed August 17, 2019, https://yesterdaysisland.com/2010/science/13-hot.php

19. W.C. **Hardwick**, "Monthly Fog Frequency in the Continental United States" *Monthly Weather Review*, (Vol. 101, pp. 763–766) Accessed August 17, 2019, https://doi.org/10.1175/1520-0493(1973)101<0763:MFFITC>2.3.CO;2

20. Sarah Oktay, "June was for Sheep," *Yesterday's Island* July 4, 2007. https://www.yesterdaysisland.com/archives/articles/sheep.php.

21. Evan Andrews, "The Sinking of the Andrea Doria" Updated 24, 2019. http://www.history.com/news/the-sinking-of-andrea-doria

22. New England Historical Society, "The Sinking of the Andrea Doria Ends an Era," Modified July 2019, http://www.newenglandhistoricalsociety.com/sinking-andrea-doria/.

23. United States Lightship Museum, Inc. "Nantucket Lightship LV-112; Crew Quotes" Published 2019. http://www.nantucketlightshiplv-112.org/crew_quotes.htm.

24. Gertrude Dunham, "Tuckernuck and the Yoho," *Proceedings of the Nantucket Historical Association,* 1921. Full text accessed digitally August 17, 2019, https://archive.org/details/proceedingsofnan00nant/page/n3.

"What Lies Beneath"

(September 12, 2013)

https://yesterdaysisland.com/lies-beneath/

25. Center for Coastal Studies in Provincetown. "Seafloor Mapping" Accessed August 16, 2022, https://coastalstudies.org/marine-geology-2/seafloor-mapping/.

26. Henry George Liddell and Robert Scott. "bathymetry" in *A Greek-English Lexicon,* Perseus Digital Library. Accessed August 16, 2019,

https://www.perseus.tufts.edu/hopper/text?doc=Perseus:text:1999.04.0057:entry=lo/gos.

27. "Sound, v2". Oxford English Dictionary (Second ed.). Oxford, England: Oxford University Press. 1969.

28. M.R. Wernand, "On the history of the Secchi disc" *Journal of the European Optical Society - Rapid publications*, Europe, v. 5, Apr. 2010. ISSN 1990-2573. https://www.jeos.org/index.php/jeos_rp/article/view/100 13s. doi:199.

29. Allen Mordica, "The Lead Line: Construction and Use" The Navy and Marine Living History Association. Accessed August 16, 2019, http://www.navyandmarine.org/planspatterns/soundingli ne.htm

30. "Leadsmen" Merriam Webster Dictionary. Accessed August 16, 2019. https://www.merriam-webster.com/dictionary/leadsman

31. Gregory Couch, "The Bonanza King" Reprinted by permission of Scribner, an Imprint of Simon & Schuster, Inc. *Time Magazine* (June 19, 2018). https://time.com/5313628/mark-twain-real-name/.

32. Mark Borrelli, Graham Giese, Stanley Dingman, Allen Gontz, M. B. Adams, A. R. Norton, T. L. B. Brown. "Linear Scour Depressions or Bedforms? Using Interferometric Sonar to Investigate Nearshore Sediment Transport," AGU Fall Meeting Abstracts. OS13B-1536, presented at 2011 Fall Meeting, AGU, San Francisco, Calif., 05-09 Dec.

33. Definitions.net, STANDS4 LLC, 2019. "insonify." Accessed August 25, 2019, https://www.definitions.net/definition/insonify.

34. Provincetown Center for Coastal Studies, "Nearshore Mapping," Accessed August 17, 2019.

http://www.coastalstudies.org/what-we-do/land-sea/nearshoremapping.htm.

35. Caleb Gostnell, Jake Yoos, and Steve Brodet, S. "NOAA Test and Evaluation of Phase Differencing Bathymetric Sonar Technology," Accessed August 25, 2019, http://citeseerx.ist.psu.edu/viewdoc/download?doi=10.1.1.534.5411&rep=rep1&type=pdf.

36. Massachusetts Department of Environmental Protection Eelgrass Mapping Viewer. Accessed on August 17, 2019. http://maps.massgis.state.ma.us/images/dep/eelgrass/eelgrass_map.htm

37. Andrea M. Caires & Sudeep Chandra. "Conversion factors as determined by relative macroinvertebrate sampling efficiencies of four common benthic grab samplers," *Journal of Freshwater Ecology*, 27:1, (2012): 97-109, DOI: 10.1080/02705060.2011.619375

38. Michael Sanders, "A Truly Miraculous Creature: Can Oysters help New England's Coast survive Climate Change?" *Yankee Magazine* Nov/Dec 2015. Accessed August 16, 2019. https://newengland.com/yankee-magazine/living/new-england-environment/oysters-3/.

"Salt Spray - Like a giant margarita"

(Sept 15, 2011) divided into two parts

http://yesterdaysisland.com/2011/science/19.php

39. National Ocean and Atmospheric Administration (NOAA) Weather Prediction Center. "Hurricanes and Extreme Rainfall," Page 1637, Accessed July 10, 2019. https://www.wpc.ncep.noaa.gov/research/mcs_web_test_test_files/Page1637.htm.

40. Richard J. Pasch, Todd B. Kimberlain and Stacy R. Stewart, (November 18, 1999). "Preliminary Report: Hurricane Floyd", National Hurricane Center. Updated Sept 9, 2014. Accessed August 25, 2019. https://www.nhc.noaa.gov/data/tcr/AL081999_Floyd.pdf
.

41. Megan E. Griffiths, "Salt Spray Accumulation and Heathland Plant Damage Associated with a Dry Tropical Storm in Southern New England," *Journal of Coastal Research* 22, no. 6 (2006): 1417-422. http://www.jstor.org/stable/30138406.

42. Scott A. Mandia, "The Long Island Express: The Great Hurricane of 1938". Updated Feb 9, 2018. https://www.sunysuffolk.edu/explore-academics/faculty-and-staff/faculty-websites/scott-mandia/38hurricane/damage_caused.html.

43. Arnold Arboretum, Harvard University, "Saltwater injury of woody plants resulting from the hurricane of September 21, 1938" *Bulletin of Popular Information* Series 4 Vol. VII November 3, 1939, Number 10. Accessed August 20, 2019, http://arnoldia.arboretum.harvard.edu/pdf/articles/1305.pdf.

44. Arboretum, "Saltwater injury of woody plants..."

45. Arnoldia, "Hurricane 'Donna' and its aftereffects to a Chatham, Massachusetts Garden" A continuation of the Bulletin of Popular Information of the Arnold Arboretum, Harvard University. Vol. 21, September 8, 1961, Number 11. Accessed August 20, 2019, http://arnoldia.arboretum.harvard.edu/pdf/articles/1571.pdf

46. Scott A. Mandia, "The Long Island..."

"The Halophytes: Salt Lovin' vegetation"

(September 15, 2011)

47. Todd Clary, "Super Serpents, Hyper Herpetologists and the Origins of Greek Aitch," Carmenta Blog, Published Feb 17, 2015, http://www.carmentablog.com/2015/02/17/super-serpents-hyper-herpetologists-and-the-origins-of-greek-aitch/.

48. United States Department of Agriculture, Natural Resources Conservation Service, "Spartina alterniflora Loisel, Plant profile," Accessed August 25, 2019. https://plants.usda.gov/core/profile?symbol=SPAL.

49. R. W. Tiner, "Wetlands of Cape Cod and the Islands, Massachusetts: Results of the National Wetlands Inventory and Landscape-level Functional Assessment," *National Wetlands Inventory report*. U.S. Fish and Wildlife Service, Northeast Region, Hadley, MA. 2010. 78 pp. plus appendices. Ref. pp 28.

50. University of Texas at Austin Lady Bird Johnson Wildflower Center, "*Baccharis halimifolia*," Modified January 13, 2016. http://www.wildflower.org/plants/result.php?id_plant=BAHA.

51. Timothy R. Van Deelen, "*Baccharis halimifolia*" In: Fire Effects Information System, [Online]. U.S. Department of Agriculture, Forest Service, Rocky Mountain Research Station, Fire Sciences Laboratory (Producer). Published 1991, Accessed August 25, 2019, https://www.fs.fed.us/database/feis/plants/shrub/bachal/all.html.

52. K. S. Bawa, "Evolution of Dioecy in Flowering Plants". *Annual Review of Ecology and Systematics*. Volume 11: 15–39. (1980):

doi:10.1146/annurev.es.11.110180.000311. JSTOR 2096901.

"Impacting the Globe – The Anthropocene"

(August 6, 2015)

https://yesterdaysisland.com/0806201512-impacting-the-globe/

53. Organization of Biological Field Stations home page. Accessed August 25, 2019, (www.obfs.org)

54. Andy Revkin, "Dot Earth," New York Times Blog. Last modified December 5, 2016. Accessed August 25, 2019. http://dotearth.blogs.nytimes.com/,

55. Andy Revkin, "Global Warming: Understanding the Forecast," Abbeville Press, March 1, 1992. 180 pages.

56. A. Will Steffen, Jacques Grinevald , Paul Crutzen and John McNeill, "The Anthropocene: conceptual and historical perspectives" *Philosophical Transactions of the Royal Society A: Mathematical, Physical and Engineering Sciences* , Vol 369 Issue 1938, Published:13 March 2011 https://doi.org/10.1098/rsta.2010.0327

57. Lynn Morris, Atlantic Rising: "Living on the edge on Nantucket Island in the US" Ecologist: The journal for the Post-industrial Age. Published September 28, 2010, http://www.theecologist.org/blogs_and_comments/Blogs /atlantic_rising/613961/atlantic_rising_living_on_the_e dge_on_nantucket_island_in_the_us.html

58. Andrew C. Revkin, "Linking Students Over Rising Atlantic Waters" Dot Earth New York Times Blogs. Published online September 27, 2010, http://dotearth.blogs.nytimes.com/2010/09/27/linking-students-over-rising-atlantic-waters .

59. Town of Nantucket Coastal Management Plan. 2014. Accessed August 25, 2019, https://nantucket-ma.gov/DocumentCenter/View/6333/Nantucket-Coastal-Management-Plan---2014

60. National Ocean and Atmospheric Administration (NOAA): NOAA Tides and Currents page "Sea level trends," Accessed March 14th, 2020. http://tidesandcurrents.noaa.gov/sltrends/sltrends.html

61. Rebecca Lindsey, National Ocean and Atmospheric Administration (NOAA): Climate Change: Global Sea Level. Published April 19, 2022. https://www.climate.gov/news-features/understanding-climate/climate-change-global-sea-level

62. Robert N. Oldale, "Coastal Erosion on Cape Cod: Some Questions and Answers". WoodsHole.er.usgs.gov. Accessed June 5, 2022 woodshole.er.usgs.gov/staffpages/boldale/capecod/quest.html

63. Fritz, Angela, "Boston's record-setting snow blitz — a winter's worth of snow in less than 10 days," Washington Post. (February 3, 2015) ISSN 0190-8286.

64. Alex Tingle, Sea Level Rise Mapping online app, Nasa data, Accessed August 16, 2022, http://flood.firetree.net/

65. National Ocean and Atmospheric Administration (NOAA) Office for Coastal Management, "Office for Coastal Management DigitalCoast," last modified: 02/10/2020. https://coast.noaa.gov/digitalcoast/

66. Commonwealth of Massachusetts, Massachusetts Office of Coastal Zone Management, "Massachusetts Sea Level Rise and Coastal Flooding Viewer." Accessed August 16, 2022. https://www.mass.gov/service-details/massachusetts-sea-level-rise-and-coastal-flooding-viewer

67. Jon Aars, Magnus Andersen, Agnes Brenière, & Samuel Blanc, "White-beaked dolphins trapped in the ice and eaten by polar bears," *Polar Research*, 34. (2015). https://doi.org/10.3402/polar.v34.26612.

"Is Nantucket Sinking?"

(May 30, 2012)

https://yesterdaysisland.com/is-nantucket-sinking/

68. Peter Reuell, "And Now, Land may be Sinking," The Harvard Gazette, February 19, 2019, https://news.harvard.edu/gazette/story/2019/02/study-of-sea-level-rise-finds-land-sinking-along-east-coast/

69. Jaap H. Nienhuis, Torbjörn E. Törnqvist, Krista L. Jankowski, Anjali M. Fernandes, and Molly E. Keogh, "A New Subsidence Map for Coastal Louisiana," *GSA Today*, v. 27, doi: 10.1130/GSATG337GW.1. Copyright 2017, The Geological Society of America, Accessed March 10, 2020. https://www.geosociety.org/gsatoday/groundwork/G337GW/GSATG337GW.pdf

70. Mary Caperton Morton, "With nowhere to hide from rising seas, Boston prepares for a wetter future," August 6, 2019, https://www.sciencenews.org/article/boston-adapting-rising-sea-level-coastal-flooding

71. National Resources Defense Council, "Water Facts: Boston, Massachusetts: Identifying and Becoming More Resilient to Impacts of Climate Change," July 2011, https://www.nrdc.org/sites/default/files/ClimateWaterFS_BostonMA.pdf.

72. "Shifting Sands" August 11, 2011 https://yesterdaysisland.com/2011/science/14.php and

"Coastal Beach Processes and Erosion" June 20, 2013 https://yesterdaysisland.com/coastal-beach-processes-erosion/

73. Rebecca Lindsey, National Oceanographic and Atmospheric Administration, "Climate Change: Global Sea Level" Published April 19, 2022. https://www.climate.gov/news-features/understanding-climate/climate-change-global-sea-level.

74. National Ocean and Atmospheric Administration (NOAA): NOAA "Tides and Currents," Accessed August 19, 2022, https://tidesandcurrents.noaa.gov/waterlevels.html?id=8443970&units=standard&bdate=19500101&edate=20161231&timezone=GMT&datum=MSL&interval=m&action=data.

75. Sea Level Rise.org. "Massachusetts' Sea Level is Rising and it is costing over 1 Billion" accessed August 19 2022, https://sealevelrise.org/states/massachusetts/.

76. John A. Church, PeterU. Clark, Anny Cazenave, Jonathan M. Gregory, Svetlana Jevrejeva, Anders Levermann, Mark A. Merrifield, Glenn A. Milne, R. Steven Nerem, Patrick D. Nunn, Antony J. Payne, W. Tad Pfeffer, Detlef Stammer and Alakkat. S. Unnikrishnan, "Sea Level Change". In: *Climate Change 2013: The Physical Science Basis. Contribution of Working Group I to the Fifth Assessment Report of the Intergovernmental Panel on Climate Change*, Cambridge University Press, Cambridge, United Kingdom and New York, NY, USA. 80 pp.

77. Morton, "with nowhere to hide..." August 6, 2019.

78. Kenneth Orris Emery, "A Simple Method of Measuring Beach Profiles," *Limnology and Oceanography*, Volume 6, (2003). doi: 10.4319/lo.1961.6.1.0090. Accessed May 3, 2020.https://seagrant.whoi.edu/wp-

content/uploads/sites/106/2015/01/Beach-and-Dune-Profiles-An-Educational.pdf

79. James F. O'Connell and Stephen P. Leatherman, "Coastal Erosion Hazards and Mapping Along the Massachusetts Shore;" *Journal of Coastal Research*, Special Issue NO. 28. Coastal Erosion Mapping and Management (Spring 1999), pp. 27-33, Published by: Coastal Education & Research Foundation, Inc. https://www.jstor.org/stable/25736182

80. Robert N. Oldale, "Coastal Erosion on Cape Cod..."

81. Seagrant Program, Woods Hole Oceanographic Institution, "Focal Point: April 2000. Shoreline Change and the importance of Coastal Erosion," Accessed April 15 2020. https://seagrant.whoi.edu/wp-content/uploads/sites/106/2015/01/WHOI-G-00-001-Sea-Grant-Shoreline-Change-a.pdf.

82. Chad Nelson, "Collecting & Using Economic Information to Guide the Management of Coastal Recreational Resources," March 2012- a doctoral dissertation the University of California Los Angeles. Accessed June 20, 2020, http://public.surfrider.org/files/nelsen/Chapter4-BeachSandMitigation.pdf (page 100).

Barrier Beaches: Nature's Fortress

June 27, 2013

https://yesterdaysisland.com/barrier-beaches-natures-fortress/

83. Nantucket Conservation Foundation, "Habitat Types; Beaches," Accessed August 1, 2022, https://www.nantucketconservation.org/properties/habitat-types/beaches/

84. The December 2008 Marine Extension Bulletin published by the Woods Hole Sea Grant & Cape Cod Cooperative Extension entitled "Coastal Dune Protection and Restoration: Using "Cape" American Beach Grass and Fencing Accessed July 10, 2022, http://www.whoi.edu/fileserver.do?id=87224&pt=2&p=8 8900

85. Nantucket Conservation Foundation, "Habitat Types: Beaches," as above.

86. United States Department of Agriculture, Natural Resources Conservation Service, "*Ammophila Breviligulata* Fernald, America beachgrass," accessed August 21,2022, https://plants.usda.gov/core/profile?symbol=AMBR

87. United States National Park Service, "National Park Service Guide to Cape Cod Beaches" Accessed July 1, 2022, http://www.nps.gov/caco/planyourvisit/upload/FinalGGC oastal.pdf.

Thar She Blows

(author's archives, Sept 7, 2011)

88. Lixion A. Avila and John Cangialosi, National Hurricane Center, "Tropical Cyclone Report Hurricane Irene (AL092011) 21-28 August 2011" 14 December 2011. Updated 19 December 2011 to correct landfall pressure in New Jersey, Updated 28 February 2012 to correct the longitude of an observation in Table 3, Updated 11 April 2013 to revise total damage estimate. Accessed June 21 2020, https://www.nhc.noaa.gov/data/tcr/AL092011_Irene.pdf.

89. Associated Press via Star News Online. Posted August 28, 2011, updated August 29, 2011. "Hurricane Irene opens new inlets on Hatteras Island," Accessed June

21, 2020,
https://www.starnewsonline.com/article/NC/20110828/N
ews/605051314/WM.

90. Jeremie, Smith, "MEMA: A Look Back at the Most
Notable Hurricanes to Hit New England" Jun 13, 2011
Massachusetts Emergency Management Agency, Accessed
June 21, 2020,
https://patch.com/massachusetts/medfield/mema-a-
look-back-at-the-most-notable-hurricanes-to-
hi12a5f21280.

91. Chris Landsea, National Ocean and Atmospheric
Administration's (NOAA) Atlantic Oceanographic and
Meteorological Research Lab, "Hurricanes: Frequently
asked Questions," Revised June 1, 2021,
http://www.aoml.noaa.gov/hrd/tcfaq/A1.htm.l

92. Ranjan Ratnakar Kelkar, "Cloud and Sunshine;
Prof. R.R. Kelkar's Blog on Weather and Climate."
Accessed on July 7, 2020,
https://rrkelkar.wordpress.com/2011/08/03/the-origin-
of-the-word-cyclone/.

93. Lexico, "typhoon," Oxford English and Spanish
Dictionary. Accessed July 7, 2020,
https://www.lexico.com/en/definition/typhoon.

94. Lexico, "Hurricane," Oxford English and Spanish
Dictionary. Accessed July 7, 2020,
https://www.lexico.com/en/definition/hurricane.

95. New World Encyclopedia entry "Tropical Cyclone"
based on Wikimedia and references therein. Accessed July
21, 2020,
https://www.newworldencyclopedia.org/entry/tropical_cy
clone.

96. Peter Adamson, "Clement Lindley Wragge and the
naming of weather disturbances," *Weather* Vol 57, Sept
2003, doi: 10.1256/wea.13.03.

The Trifecta of Moon Gazing Nights

(October 1, 2015)

https://yesterdaysisland.com/1001201512-the-trifecta-of-moon-gazing-nights/

97. Clyde Simpson, "Everything Ohio observers need to know," Cleveland Museum of Natural History, Accessed July 31 2022, https://www.cmnh.org/science-news/blog/september-2015/total-lunar-eclipse-september-27

98. Clyde Simpson, "Everything Ohio…"

99. Sarah Oktay, "Living with the Sea" *Yesterday's Island*. November 2009. http://yesterdaysisland.com/archives/science/21.php.

100. Florida Fish and Wildlife Conservation Commission. "Artificial Lighting and Sea Turtle Behavior" Accessed August 16, 2022, https://myfwc.com/research/wildlife/sea-turtles/threats/artificial-lighting/.

101. Alexa Stevenson, "Probing Question: Why are moths attracted to light?" Penn State (October 20, 2008) Accessed August 16, 2022. http://news.psu.edu/story/141283/2008/10/20/research/probing-question-why-are-moths-attracted-light.

102. Jim McCormac, "Night Flowering Primrose," Ohio Birds and Biodiversity, Accessed August 16, 2022., http://jimmccormac.blogspot.com/2011/07/night-flowering-primrose.html.

Restoring Our Marshes

(July 31-August 6, 2008)

https://yesterdaysisland.com/2008/features/14.php

103. Tod Highsmith, "9 Wetland Monsters from World Folklore," Wisconsin Wetlands Association, Oct 25, 2017. Accessed July 21, 2020, https://www.wisconsinwetlands.org/updates/9-wetland-monsters/

104. Massachusetts Open Marsh Water Management Workgroup, May 2010, "Massachusetts Mosquito Control, Open Marsh Water Management Standards", accessed July 21, 2020 at https://www.mass.gov/doc/open-water-marsh-management-standards/download

105. Mass OMWMW, May 2010... as above.

106. John Tibbetts, "Louisiana's wetlands: a lesson in nature appreciation." *Environmental Health Perspectives* Vol. 114,1 (2006): A40-3. doi:10.1289/ehp.114-a40

107. Christine C. Shepard, Caitlin Crain, and Michael W. Beck, "The protective role of coastal marshes: a systematic review and meta-analysis," *PloS One* vol. 6,11 (2011): e27374. doi:10.1371/journal.pone.0027374

Section 2: It's a Hard-Shelled Life

What the heck is monkey dunk?

(August 8, 2013)

https://yesterdaysisland.com/sponges-climate-change-bay-scallops/

108. Ludwig-Maximilians-Universitaet Muenchen (LMU). "Animal evolution: Sponges really are oldest animal phylum." ScienceDaily. Accessed August 8, 2020, www.sciencedaily.com/releases/2015/12/151201130729.htm.

109. Hestetun Jon Thomassen, Rapp Hans Tore, Pomponi Shirley, "Deep-Sea Carnivorous Sponges from the Mariana Islands," *Frontiers in Marine Science* Volume 6 (2019) https://www.frontiersin.org/article/10.3389/fmars.2019.00371   DOI=10.3389/fmars.2019.00371

110. Kristen Wheeler, "Demospongiae," Animal Diversity Web, University of Michigan Museum of Zoology, accessed July 21, 2020, https://animaldiversity.org/accounts/Demospongiae/.

111. Kristen Wheeler, "Demospongiae," as above.

112. World Porifera Database, Accessed August 9, 2020, http://www.marinespecies.org/porifera/porifera.php?p=taxdetails&id=134121.

113. Richard E. Grant, "Notice of a New Zoophyte (*Cliona celata* Gr.) from the Firth of Forth" (1826) Edinburgh New Philosophical Journal 1: 78-81.

114. John M. Guinotte and Victoria J. Fabry, "Ocean acidification and its potential effects on marine ecosystems" *Ann N Y Acad Sci* 1134. (2008): 320–342.

115. Alan R. Duckworth and Bradley J. Peterson, "Effects of seawater temperature and pH on the boring rates of the sponge **Cliona celata** in scallop shells" *Marine Biology* September 14 2012, http://blueocean.org/documents/2012/10/marine-sponges-bore-into-shellfish-faster-due-to-effects-of-climate-change.pdf.

116. The Royal Society. Ocean acidification due to increasing atmospheric carbon dioxide. *The Royal Society.* (2005). Accessed August 9, 2020, http://www.us-ocb.org/publications/Royal_Soc_OA.pdf.

117. Alan R. Duckworth and Bradley J. Peterson, as above.

Hermaphrodites in the Harbor

(May 20, 2010)

https://yesterdaysisland.com/2010/science/3-hermaph.php

118. Encyclopedia of Life. "Common Slipper Shell, Accessed August 15, 2020, https://eol.org/pages/593855/articles.

119. Serge Gofas, *Crepidula fornicata* (Linnaeus, 1758). *World Marine Mollusca database.* Accessed August 15, 2020: World Register of Marine Species at http://www.marinespecies.org/aphia.php?p=taxdetails&id=138963.

120. Thanumalaya Subramoniam, "Sexual Biology and Reproduction in Crustaceans," Academic Press 2017. Pages 57-103, https://doi.org/10.1016/B978-0-12-809337-5.00003-4

121. Serge Gofas, as above.

122. Smithsonian Environmental Research Center, National Exotic Marine and Estuarine Species Information (NEMESIS) database. *Codium fragile ssp. Fragile* Accessed August 16, 2022. https://invasions.si.edu/nemesis/browseDB/SpeciesSummary.jsp?TSN=6897

123. Abigail E. Cahill, Alia Rehana Juman, Aaron Pellman-Isaacs, and William T. Bruno. "Physical and Chemical Interactions with Conspecifics Mediate Sex Change in a Protandrous Gastropod *Crepidula fornicata.*" *Biol Bull.* 229(3) (2015):276-81.

124. Matthew J. Harke, Christopher J. Gobler, and Sandra E. Shumway, "Suspension feeding by the Atlantic slipper limpet (*Crepidula fornicata*) and the northern quahog (Mercenaria mercenaria) in the presence of

cultured and wild populations of the harmful brown tide alga, *Aureococcus anophagefferens*" *Harmful Algae* 10, no. 5 (2011): 503-511. doi: 10.1016/j.hal.2011.03.005

125. Gourmet Pedia "Slipper Limpet" recipe, Accessed August 15, 2020, http://gourmetpedia.net/products/shellfish/slipper-limpet/

126. Sara P. Grady, Deborah Rutecki, Ruth Carmichael, and Ivan Valiela, "Age Structure of the Pleasant Bay Population of *Crepidula fornicata*: A Possible Tool for Estimating Horseshoe Crab Age," *Biol. Bull.* 201: (October 2001): 296–297.

Feeling Crabby?

(Sept 2-8, 2010)

https://yesterdaysisland.com/2010/science/18.php

127. Sammy De Grave, N. Dean Pentcheff, Shane T. Ahyong, et al., "A classification of living and fossil genera of decapod crustaceans". *Raffles Bulletin of Zoology*. Suppl. 21: (2009). 1–109.
128. Henry George Liddell and Robert Scott, (1940) A Greek–English Lexicon, Oxford: Clarendon Press
129. Nola Taylor Redd, "The Crab Nebula (M1): Facts, Discovery & Images" Space.com. Accessed August 16, 2020, https://www.space.com/16989-crab-nebula-m1.html
130. Mandal, Ananya, "Cancer History," News-Medical. Accessed August 16, 2020, https://www.news-medical.net/health/Cancer-History.aspx.
131. George Gordh, Gordon Gordh, & David Headrick. A Dictionary of Entomology, CAB International, (2003) p. 182. ISBN 978-0-85199-655-4.

132. Catherine E. Derivera, Gregory M. Ruiz, Anson H. Hines, and Paul Jivoff, "Biotic Resistance to Invasion: Native Predator Limits Abundance and Distribution of an Introduced Crab"; *Ecology.* 86 (2005):3364–3376. http://dx.doi.org/10.1890/05-0479

133. *The Blue Crab Archives,* Hosted by ZINSKINET, LLC. Accessed November 28, 2020. https://www.bluecrab.info/mating.html

134. The Blue Crab Archives as above.

135. The Blue Crab Archives as above.

136. *The Blue Crab Archives,* Hosted by ZINSKINET, LLC. Accessed Sept 1st, 2021, https://www.bluecrab.info/lifecycle.html

137. Hilary Lane Glandon, Hali Kilbourne, & Thomas J. Miller, "Winter is (not) coming: Warming temperatures will affect the overwinter behavior and survival of blue crab," *PloS One,* 14(7), e0219555. (2019), https://doi.org/10.1371/journal.pone.0219555

138. Stephen Tomasetti, Brook K. Morrell, Lucas R. Merlo, ad Christopher J. Gobler, "Individual and combined effects of low dissolved oxygen and low pH on survival of early stage larval blue. crabs, *Callinectes sapidus,*" *PLoS ONE 13*(12): e0208629. (2018), https://doi.org/10.1371/journal.pone.0208629

139. https://www.bayjournal.com/news/fisheries/survey-yields-mixed-verdict-on-blue-crab-stock/article_fe3fe718-bc86-11eb-9ecd-c7d5afd8cc54.html

140. David Samuel Johnson, "The Savory Swimmer Swims North: A Northern Range Extension of the Blue Crab *Callinectes Sapidus?,*" *Journal of Crustacean Biology,* Volume 35, Issue 1, 1 January 2015, Pages 105–110, https://doi.org/10.1163/1937240X-00002293

141. Chesapeake Bay Program Field guides: Common spider crab: *Libinia emerginata* https://www.chesapeakebay.net/S=0/fieldguide/critter/common_spider_crab accessed July 17 2022.

142. Gertrude W. Hinsch, "Reproductive behavior in the spider crab, *Libinia emarginata* (L.)". *The Biological Bulletin*. 135 (2): 273–278. (1968). doi:10.2307/1539781. JSTOR 1539781.

143. Hinsch, "Reproductive behavior..."273-278.

144. John C. Aldrich, "The spider crab *Libinia emarginata* Leach, 1815 (Decapoda Brachyura), and the starfish, an unsuitable predator but a cooperative prey". *Crustaceana*. 31 (2): 151–156. (1976), doi:10.1163/156854076X00189. JSTOR 20103088.

145. Chesapeake Bay Program Field guides, "Common spider crab: *Libinia emerginata,*" Accessed July 17 2022. https://www.chesapeakebay.net/S=0/fieldguide/critter/common_spider_crab.

146. Chesapeake Bay, "Common Spider crab".

147. iNaturalist, Longnose Spider Crab: *Libinia dubia,*" Accessed August 21, 2022, https://www.inaturalist.org/guide_taxa/255118

148. David J. Stasek, Jeffrey E. Tailer, James Page, Patrick J. Geer, and Bryan A. Fluech "The Nature of the Symbiosis between Cannonball Jellyfish and Spider Crabs in Georgia's Coastal Waters," *Southeastern Naturalist* 19(2), (28 April 2020): 233-240. https://doi.org/10.1656/058.019.0204

The Beautiful and Vicious Lady Crab

(August 27, 2015)

https://yesterdaysisland.com/0827201512-the-beautiful-and-vicious-lady-crab/

149. Linda L. Stehlik, "Diets of the Brachyuran Crabs Cancer *Irroratus*, *C. Borealis*, and *Ovalipes Ocellatus* in the New York Bight." *Journal of Crustacean*

Biology, vol. 13, no. 4, 1993, pp. 723–35. JSTOR, https://doi.org/10.2307/1549103.

150. Integrated Taxonomic Information System (ITIS) on-line database, *"Ovalipes ocellatus,"* Accessed 17 Jul. 2022, https://www.itis.gov/servlet/SingleRpt/SingleRpt?search_topic=TSN&search_value=98714#null.

151. Mary Jane Rathbun, "The Brachyura collected by the U. S. Fish Commission steamer Albatross on the voyage from Norfolk, Virginia, to San Francisco, California, 1887-1888". Proceedings of the United States National Museum. 21 (1162): 567–616. (1898), doi:10.5479/si.00963801.21-1162.567.

152. Eugene H. Kaplan, "Lady crab *Ovalipes ocellatus*". In Roger Tory Peterson (ed.). A Field Guide to Southeastern and Caribbean Seashores: Cape Hatteras to the Gulf Coast, Florida, and the Caribbean. Peterson Field Guides (2nd ed.). Houghton Mifflin Harcourt. (1999). p. 322. ISBN 978-0-395-97516-9.

153. Stephan Gregory Bullard, *"Ovalipes ocellatus* (Herbst, 1799)"*. Larvae of anomuran and brachyuran crabs of North Carolina: a guide to the described larval stages of anomuran (families Porcellanidae, Albuneidae, and Hippidae) and brachyuran crabs of North Carolina, U.S.A. Volume 1 of Crustaceana monographs. Brill. (2003). pp. 29–30.

154. Linda L. Stehlik, "Diets of the Brachyuran Crabs..."

155. University of Rhode Island Extension service. Adapted from The Uncommon Guide to Common Life on Narragansett Bay. Save The Bay, 1998, accessed July 17 2022, http://www.edc.uri.edu/restoration/html/gallery/invert/lady.htm

156. Eugene H. Kaplan, "Lady crab..."

Beach Houdinis: Mole Crabs

(May 21-27, 2009)

https://yesterdaysisland.com/archives/science/3.php

157.   Ioannis Ant. Scopoli (1777). Introductio ad historiam naturalem sistens genera lapidum, plantarum, et animalium hactenus detecta, caracteribus essentialibus donata, in tribus divisa, subinde ad leges naturae (in Latin). Prague: Wolfgang Gerle.

158.   Say, T. (1817-1818). An account of the Crustacea of the United States. *Journal of the Academy of Natural Sciences of Philadelphia.* 1(1):57-63, 65-80 (pl.), 97-101, 155-169 (1817); 1(2): 235-253, 313-319, 374-401, 423-441 (1818). https://www.biodiversitylibrary.org/part/2444 01#/summary.

159.   Dorothy A. Horn, D.A., Elise F. Granek, and Clare L. Steele, Effects of environmentally relevant concentrations of microplastic fibers on Pacific mole crab (*Emerita analoga*) mortality and reproduction. *Limnol Oceanogr Lett*, 5: (2020): 74-83. https://doi.org/10.1002/lol2.10137

160.   Richard B. Forward, Jr. Humberto Diaz, & Jonathan Cohen, "The tidal rhythm in activity of the mole crab *Emerita talpoida,*" *Journal of the Marine Biological Association of the United Kingdom* 85. 895 - 901. (2005). 10.1017/S0025315405011860.

Ghost Crabs

(August 1, 2013)

https://yesterdaysisland.com/ghost-crabs/

161.   George Karleskint, Richard Keith Turner, and James Small, "Intertidal communities," Introduction to Marine Biology (3rd ed.). Cengage Learning. (2009): pp. 356–411.

162.   Andrew M. Phillips, "The ghost crab – adventures investigating the life of a curious and interesting creature that lives on our doorstep, the only large

crustacean of our North Atlantic coast that passes a good part of its life on land," *Natural History* (1940): 46:36-41.

163. George Karleskint, Richard Keith Turner, and James Small. "Intertidal communities," *Introduction to Marine Biology* (3rd ed.). Cengage Learning. (2009): pp. 356–411.

164. Lorus J. Milne and Margery J. Milne. "Notes on the Behavior of the Ghost Crab." The American Naturalist 80, no. 792 (1946): 362–80. http://www.jstor.org/stable/2457854.

165. Keith Davey, "Ocypode cordimana (Family Ocypodidae)". SpeciesBank. Department of the Environment, Water, Heritage and the Arts. Accessed August 6, 2022, http://www.environment.gov.au/cgi-bin/species-bank/sbank-treatment.pl?id=77373.

166. Joaquim Branco, Juliano Hillesheim, Helio Fracasso, Martin Christoffersen, Cristiano Evangelista. "Bioecology of the ghost crab Ocypode quadrata (Fabricius, 1787) (Crustacea: Brachyura) compared with other intertidal crabs in the Southwestern Atlantic." *Journal of Shellfish Research,* 29 (2): (2010) 503-512.

167. Joaquim Branco et. al, "Bioecology of ghost crab" 503-512.

168. Wayne and Martha McAlister's book "Life on Matagorda Island Page 56 (Texas A&M University Press, 2004) 244 pages. [http://books.google.com/books/about/Life_on_Matagorda_Island.html?id=rOkC qL4rrHoC accessed August 6, 2022- page 56)

169. Joaquim Branco et al, "Bioecology of ghost crab" 503-512.

170. Biodiversity Works, "Ghost Crabs on Martha's Vineyard", (accessed July 28th 2013) http://biodiversityworksmv.org/atlantic-ghost-crabs-on-marthas-vineyard/

171. Mark Allen Lovewell, "Ghost Crab Arrivals May Signal Climate Shift, Threaten Plovers," Vineyard Gazette, published May 31 2012, http://mvgazette.com/news/2012/05/31/ghost-crab-arrivals-may-signal-climate-shift-threaten-plovers.

172. Charles H. Peterson, Darren H.M. Hickerson and Gina Grissom Johnson, "Short-term consequences of nourishment and bulldozing on the dominant large invertebrates of a sandy beach," *Journal of Coastal Research*. 16: (2000) 368-378.

Neighborhood Bullies, the green crab

(July 24-30 2008)

http://yesterdaysisland.com/2008/features/bullies.php

173. Constantine S. Rafinesque, "Synopsis of four new genera and ten new species of Crustacea, found in the United States," *American Monthly Magazine* 2: (1817) 40-43

174. Edwin D. Grosholz and Gregory M. Ruiz, "Predicting the impact of introduced marine species: Lessons from the multiple invasions of the European green crab *Carcinus maenas*," *Biological Conservation*, Volume 78, Issues 1–2, 1996, Pages 59-66, https://doi.org/10.1016/0006-3207(94)00018-2.

175. Greg Klassen and Andrea Locke, "A biological synopsis of the European Green Crab, *Carcinus maenas*," *Canadian Manuscript Report of Fisheries and Aquatic Sciences* 2818 (2007): 1-75

176. Nature, "America's Least Wanted", S24, EP 9, June 10, 2008, Accessed August 22,2022, https://www.pbs.org/wnet/nature/animals-behaving-worse-americas-least-wanted/911/

177. Sylvia Behrens Yamada and Laura Hauck, "Field identification of the European green crab species: *Carcinus maenas* and *Carcinus aestuarii*," Journal of Shellfish Research. 20 (3) (2001): 905–909.

178.   Steven Thomas Tettelbach, "Dynamics of Crustacean Predation on the North Bay Scallop *Argopecten irradians irradians* (Crabs Pelecypods, populations, Connecticut" (1986). *Doctoral Dissertations*. AAI8622931. https://opencommons.uconn.edu/dissertations/AAI86 22931,

179.   Andrew Cohen, James T. Carlton, and Michael. C. Fountain, "Introduction, dispersal and potential impacts of the green crab *Carcinus maenas* in San Francisco Bay, California". *Marine Biology*. 122 (2) (1995): 225–237.

180.   United States Geological Survey, "They're Healthier: With Fewer Parasites, Invaders Gain Competitive Edge Over Native Species," *ScienceDaily*, (accessed August 5, 2022), www.sciencedaily.com/releases/2003/02/0302070705 48.htm.

181.   International Maritime Organization. "Ballast Water Management" accessed August 5, 2022, https://www.imo.org/en/OurWork/Environment/Page s/BallastWaterManagement.aspx.

182.   Jukka Sassi, Satu Viitasalo, Jorma Rytkönen, & Erkki Leppäkoski, "Experiments with ultraviolet light, ultrasound and ozone technologies for onboard ballast water treatment," *VTT Research Notes*, 2313. (2006): 1-86.

183.   Margaret Osbourne, New Hampshire Distillery Makes Whiskey Out of Invasive Crabs. *Smithsonian Magazine*, July 13, 2022. https://www.smithsonianmag.com/smart-news/new-hampshire-distillery-makes-whiskey-out-of-invasive-crabs-180980414/ accessed August 7, 2022.

184.   Bouhee Kang, B., Angela D. Myracle, & Denise Skonberg, "Potential of recovered proteins from invasive green crabs (*Carcinus maenas*) as a functional food ingredient." *Journal of the Science of Food and Agriculture*, 99(4), (2018): 1748-1754.

185. Michael Fincham. "An endless invasion? Green crabs, New England intruders, move west." *Maryland Marine Notes* March-April 1996, 14(2). http://www.mdsg.umd.edu/MarineNotes/Mar-Apr96/

186. William Cameron Walton, "Attempts at physical control of Carcinus maenas within coastal ponds of Martha's Vineyard, MA (northeastern coast of North America)," *Proceedings of the first international workshop on the demography, impacts and management of introduced populations of the European crab, Carcinus maenas: 20-21 March 1997.* Centre for Research on Introduced Marine Pests, Hobart, Tasmania, Australia. Technical Report 11.

187. PIE-Rivers partnership, "Green Crab Removal," accessed August 7, 2022, https://pie-rivers.org/portfolio-item/id_31/

188. Catherine E. Derivera, Gregory M. Ruiz, Anson H. Hines and Paul Jivoff, "Biotic Resistance to Invasion: Native Predator Limits Abundance and Distribution of an Introduced Crab," *Ecology* 86 (2005): 3364–3376. http://dx.doi.org/10.1890/05-0479

189. James A. MacDonald, Ross Roudez, Terry Glover, and Judith Weis, "The invasive green crab and Japanese shore crab: behavioral interactions with a native crab species, the blue crab," *Biol Invasions* 9, (2007): 837–848. https://doi.org/10.1007/s10530-006-9085-6

190. Jens T. Høeg, N. Murphy, C. Wittwer, "Developing the options for managing marine pests: specificity trials on the parasitic castrator, *Sacculina carcini*, against the European crab, *Carcinus maenas*, and related species," *Journal of Experimental Marine Biology and Ecology,* 254 (1) (2000): 37-51.

191. Charles E. Epifanio, Invasion biology of the Asian shore crab *Hemigrapsus sanguineus*: A review. *Journal of Experimental Marine Biology and Ecology,* Volume 441, 2013, Pages 33-49, https://doi.org/10.1016/j.jembe.2013.01.010.

192. Washington Department of Fish and Wildlife website. Accessed August 7, 2022. https://wdfw.wa.gov/species-habitats/invasive/carcinus-maenas

Stealing Home

(August 27-Sept 2, 2009)

https://yesterdaysisland.com/archives/science/17.php

193. R. Huys, "An Updated Classification of the Recent Crustacea," Journal of Crustacean Biology. 23. (2009) 495-497. doi.10.1651/0278-0372(2003)023[0495:BR]2.0.CO;2.
194. Lancelot Alexander Borradaile, "Crustacea. Part II. Porcellanopagurus: an instance of carcinization". British Antarctic ("Terra Nova") Expedition, 1910–1913. Natural History Report. *Zoology*. British Museum. 3 (3) (1916).: 111–126. OCLC 1027015098
195. Harper Douglas, "Etymology of hermit," Online Etymology Dictionary, accessed August 12, 2022, https://www.etymonline.com/word/hermit.
196. Land Hermit Crab Owners Society (February 21, 2013). "How old is my hermit crab?" *The Crabstreet Journal*, accessed August 12, 2022, https://crabstreetjournal.org/blog/2013/02/21/how-old-is-my-hermit-crab/.
197. Geological Society of America. "Hermit Arthropods 500 Million Years Ago?" ScienceDaily, accessed August 11, 2022, www.sciencedaily.com/releases/2009/03/090331183512.htm.
198. Patsy A. McLaughlin, Hermit Crabs—are They Really Polyphyletic?, Journal of *Crustacean Biology*, Volume 3, Issue 4, 1 October 1983, Pages 608–621, https://doi.org/10.1163/193724083X00274
199. Squires, H.J. "Larvae of the hermit crab, *Pagurus arcuatus*, from the plankton (Crustacea, Decapoda)".

Journal of Northwest Atlantic Fishery Science 18: (1996): 43–56. doi:10.2960/J.v18.a3.

200.   Thomas Hunt Morgan (1900). Further experiments on regeneration of the appendages of the hermit crab. *Anat. Anz.* 17, 1–9.

201.   Zachary Emberts, Christine Whitney Miller, Daniel Kiehl, Colette M. St Mary, "Cut your losses: self-amputation of injured limbs increases survival," *Behav Ecol.* 2017 Jul-Aug;28(4):1047-1054. doi: 10.1093/beheco/arx063. Epub 2017 Apr 22. PMID: 29622925; PMCID: PMC5873245.

202.   Randi D. Rotjan; Jeffrey R. Chabot; Sara M. Lewis (2010). "Social context of shell acquisition in Coenobita clypeatus hermit crabs". Behavioral Ecology. 21 (3): 639–646. doi:10.1093/beheco/arq027. hdl:10.1093/beheco/arq027. ISSN 1465-7279.

203.   Ferris Jabr. "On a Tiny Caribbean Island, Hermit Crabs Form Sophisticated Social Networks," *Scientific American.* (5 June 2012) Accessed August 22, 2022. https://www.scientificamerican.com/article/vacancy-hermit-crab-social-networks/.

204.   Mark E. Laidre, "Niche construction drives social dependence in hermit crabs," *Current Biology* 22 (2012): R861- R863.

205.   Philippe de Vosjoli, "The Care of Land Hermit Crabs," Advanced Vivarium Systems; (August 1999) Accessed August 12, 2022, https://crabstreetjournal.org/blog/2012/11/03/chirping-or-croaking/.

206.   David Carlon and Ebersole, John, "Life-History Variation Among Three Temperate Hermit Crabs: The Importance of Size in Reproductive Strategies," *Biological Bulletin,* 188. (1995). doi.10.2307/1542309.

Counting "Living Fossils" (June 11 - 17, 2009) merged with "Dinosaurs Among Us" (June 9-15, 2011)

https://yesterdaysisland.com/2011/science/5.php

207. Natalie Angier, "Tallying the Toll on an Elder of the Sea". The New York Times, (June 10, 2008). Accessed August 12.2022, https://www.nytimes.com/2008/06/10/science/10ang i.html

208. Michael Ruggiero, Dennis P. Gordon, Thomas M. Orrell, Nicolas Bailly, et al., "A Higher-Level Classification of All Living Organisms," *PLoS ONE* 10(4): e0119248. (2015) doi: 10.1371/journal.pone.0119248.

209. Charles H. Calisher, "Taxonomy: what's in a name? Doesn't a rose by any other name smell as sweet?" *Croat Med J.* (2007) Apr;48(2):268-70.

210. Leif Størmer, "Phylogeny and Taxonomy of Fossil Horseshoe Crabs." Journal of Paleontology 26, no. 4 (1952): 630–40. http://www.jstor.org/stable/1299851.

211. Carl N. Shuster, Jr. "A pictorial review of the natural history and ecology of the horseshoe crab Limulus polyphemus, with reference to other Limulidae. In: Bonaventura J, Bonaventura C, Tesh S (eds) *Physiology and biology of horseshoe crabs*. Alan R. Liss, New York, (1982): pp 1–52

212. Angier, "Tallying…" NYT.

213. Schuster, (1982) A Pictorial….

214. Elizabth A. Walls, Jim Berkson, and Stephen A, Smith, "The Horseshoe Crab, *Limulus polyphemus*: 200 Million Years of Existence, 100 Years of Study," *Reviews in Fisheries Science,* 10(1) (2002): 39-73.

215. US Fish and Wildlife Service, "The Horseshoe Crab: Limulus polyphemus A Living Fossil," Accessed August 12, 2022, https://www.fws.gov/sites/default/files/documents/H orseshoecrab.pdf

216. Frederik Barry Bang. "The toxic effect of a marine bacterium on Limulus and the formation of blood clots," *Biol Bull* 105 (1953):447-448.

217. Jack Levin and Frederik Barry Bang, "A description of cellular coagulation in Limulus," *Bull. of Johns Hopkins Hosp.* 115(1964):337.

218. Jack Levin and Frederik Barry Bang, The role of endotoxin in the extracellular coagulation of Limulus blood. *Bull. Johns Hopkins Hosp.* 115 (1964):265.6.

219. Jack Levin and Frederik Barry Bang, "Clottable protein in Limulus: Its localization and kinetics of its coagulation by endotoxin," *Thromb Diathes Haemorrh* (Stuttg) 19 (1968):186.

220. Thomas J. Novitsky. "Biomedical implications for managing the Limulus polyphemus harvest along the northeast coast of the United States," In: Carmichael RH, Botton M, Shin PKS, Cheung SG (eds) *Changing global perspectives on horseshoe crab biology, conservation and management.* Springer, Cham, (2015) pp 483–500

221. Mike Olson, "How the blood of the horseshoe crab helps keep you alive," Wired, August 2011. https://www.wired.com/2011/11/in-evolutions-race-horseshoe-crabs-took-a-slower-pace/

222. Kevin Gilyear. "Prime Directive: An Exclusive Interview with Peter Cullen," The Digital Fix (August 2006) accessed August 12, 2022. https://www.thedigitalfix.com/film/prime-directive-an-exclusive-interview-with-peter-cullen/

223. Valaquen. Strange Shapes, 2012. Facehugger/Chestburster. https://alienseries.wordpress.com/2012/10/26/facehuggerchestburster/ accessed August 12, 2022

224. Laura Tangley. National Wildlife Federation. Shorebirds' Fate Hinges on Horseshoe Crabs. 2013. https://www.nwf.org/Magazines/National-Wildlife/2013/OctNov/Conservation/Shorebirds-and-Horseshoe-Crabs accessed August 12,2022.

225. Glenn Gauvry, "Horseshoe Crabs and Their Neighbors."Horseshoecrab.org, 2003, www.horseshoecrab.org/nh/hist.html accessed on

August 12,2022.
https://www.delawarebayhscsurvey.org/importance

226. Sarah M. Karpanty, James D. Fraser, Jim Berkson, Lawrence J. Niles, Amanda Dey, and Eric P. Smith. "Horseshoe Crab Eggs Determine Red Knot Distribution in Delaware Bay." *The Journal of Wildlife Management* 70, no. 6 (2006): 1704–10. http://www.jstor.org/stable/4128104.

227. Schuster, (1982) A Pictorial....

228. The Horseshoe Crab. Horseshoe Crabs and their Neighbors. https://www.horseshoecrab.org/nh/eco.html accessed August 12, 2022.

229. The Horseshoe Crab. Horseshoe Crab Conservation Network. https://horseshoecrab.org/conservation/ accessed August 12, 2022

230. The Horseshoe Crab. Horseshoe Crab Conservation Network.

Strange & Unusual (Sea) Creatures [partial]

(August 20-26, 2009)

https://yesterdaysisland.com/archives/science/16.php

231. Gail V. Ashton. "*Caprella mutica* Schurin, 1935". Caprellids, LifeDesks. https://web.archive.org/web/20150104184836/http:// caprellids.lifedesks.org/pages/84 Accessed August 12,2022.

232. Encarta Reference Library Premium 2005 DVD. Article - Skeleton Shrimp

233. Peter J. Bryant. "Natural History of Skeleton Shrimp," accessed August 12,2022, https://nathistoc.bio.uci.edu/crustacea/Amphipoda/C aprella.htm.

234. Monaca Noble, "Unwanted Species: The Fouling Community," Smithsonian Environmental Research

Center, Marine Invasions Research, July 1, 2012. https://serc.si.edu/labs/marine-invasions-research/feature-story/unwanted-species-fouling-community.

235.   Karin Boos, Gail V. Ashton, and Elizabeth J. Cook, "The Japanese skeleton shrimp *Caprella mutica* (Crustacea, Amphipoda): a global invader of coastal waters." *In the wrong place-alien marine crustaceans: distribution, biology and impacts* (2011) (pp. 129-156). Springer, Dordrecht.

236.   Michelle Uhr. Salen State University. B.S. Honors Thesis. "An Examination of Marine Fouling Organisms' Presence On Varying Substrates In A New England Marina," Accessed August 12, 2022, http://hdl.handle.net/20.500.13013/632

237.   Michelle Marraffini, Gail Ashton, Christopher Brown, Andrew Chang, & Gregory Ruiz, "Settlement plates as monitoring devices for non-indigenous species in marine fouling communities," *Management of Biological Invasions* 8. (2017): 559-566. 10.3391/mbi.2017.8.4.11.

Small Monsters in the Water – Chimeras

(August 16, 2012)

https://yesterdaysisland.com/small-monsters-in-the-water-chimeras/

238.   Britannica, T. Editors of Encyclopaedia, "Chimera." Encyclopedia Britannica, February 14, 2020, https://www.britannica.com/topic/Chimera-Greek-mythology.

239.   Aaron Norton and Ozzie Zehner. "Which Half Is Mommy?: Tetragametic Chimerism and Trans-Subjectivity". Women's Studies Quarterly. Fall/Winter (3–4) (2008): 106–127. doi:10.1353/wsq.0.0115

240. Corinna N. Ross, Jeffrey A. French, and Guillermo Orti. "Germ-line chimerism and paternal care in marmosets (Callithrix kuhlii)". *Proceedings of the National Academy of Sciences.* Vol. 104 (15) (2007): 6278–6282. Bibcode:2007 PNAS..104.6278R. doi:10.1073/pnas.0607426104.

241. Theodore Pietsch. "Dimorphism, parasitism and sex revisited: modes of reproduction among deep-sea ceratioid anglerfishes (Teleostei: Lophiiformes)," *Ichthyol Res.* 52 (2005):207–236.

242. Leland W. Pollock. A Practical Guide to the Marine Animals of Northeastern North America. Figure 15.19. Page 116. Rutgers University Press, 1998 - Nature - 367 pages.

243. Katherine Yi, "*Squilla empusa*," Animal Diversity Web, University of Michigan Museum of Zoology, (2001), Accessed August 14, 2022, https://animaldiversity.org/accounts/Squilla_empusa/ .

244. William James Heitler, Kayleigh Fraser and Enrico A. Ferrero. "Escape behaviour in the stomatopod crustacean squilla mantis, and the evolution of the caridoid escape reaction," *Journal of Experimental Biology,* v203 (Jan 15 2000): 183-192. DOI: 10.1242/jeb.203.2.183

245. Ed Yong. "The Mantis Shrimp – the world's fastest punch," Not Exactly Rocket Science, accessed August 12th, 2022, http://notexactlyrocketscience.wordpress.com/2006/08/28/mantis-shrimps-the-worlds-fastest-punch/

246. Sheila N. Patek, Wyatt L. Korff, & Roy L. Caldwell. "Biomechanics: deadly strike mechanism of a mantis shrimp." *Nature*, 428(6985), (2004) 819–820. https://doi.org/10.1038/428819a

247. Timothy York, Samuel B. Powell, Shengkui Gao, Lindsey Kahan, Tauseef Charanya Debajit Saha, Nicholas W Roberts, Thomas W. Cronin; Justin Marshall, Samuel Achilefu, Spencer P. Lake,

Baranidharan Raman, and Viktor Gruev (2014). "Bioinspired polarization imaging sensors: from circuits and optics to signal processing algorithms and biomedical applications". Proceedings of the IEEE. 102 (10): 1450–1469. doi:10.1109/JPROC.2014.2342537. PMC 4629637. PMID 26538682.

248.   Thomas W. Cronin, Justin Marshall, Michael F. Land, "The unique visual system of the mantis shrimp," *American Scientist*, v82 (1994): 356-366.

249.   Gabriella Hannah Wolff, Hanne Halkinrud Thoen, Justin Marshall, Marcel E Sayre, Nicholas James Strausfeld, "An insect-like mushroom body in a crustacean brain," *eLife* 6:e29889. (2017) https://doi.org/10.7554/eLife.29889

250.   Nick Dakin. The Marine Aquarium. London: Andromeda. (2004) ISBN 978-1-902389-67-7.

Floating Wonders (Aug 28-Sept 3, 2008) combined with "Blue Bottle Washashore" (June 13, 2013)

http://www.yesterdaysisland.com/2008/features/18c.php

https://yesterdaysisland.com/bluebottle-washashore-portuguese-man-o-war/

251.   Biology online, "Cnidocyte," accessed August 14,2022, https://www.biologyonline.com/dictionary/cnidocyte

252.   Gabriele Kass-Simon and A.A. Scappaticci Jr, "The behavioral and developmental physiology of nematocysts," *Canadian Journal of Zoology*, 80 (10) (2002): 1772–1794. doi:10.1139/Z02-135

253.   Leslie S. Babonis, Camille Enjolras, Joseph F. Ryan and Mark Q. Martindale, "A novel regulatory gene promotes novel cell fate by suppressing ancestral fate in the sea anemone *Nematostella vectensis*," *PNAS* May

2, 2022 Volume 119 (19)
https://doi.org/10.1073/pnas.2113701119

254. Red Cross, "Health Safety Advisory: Jellyfish Stings," accessed August 14, 2022, https://www.redcross.org/content/dam/redcross/Health-Safety-Services/scientific-advisory-council/Scientific%20Advisory%20Council%20ADVISORY%20-%20Jellyfish%20Stings.pdf.

255. Kim Carollo, ABC News. "Jellyfish Stings: Urine and Other Remedies May Not Work" (2012), accessed August 14, 2022, http://abcnews.go.com/Health/jellyfish-stings-urine-effective-remedy/story?id=16630841#.UB-OlJGnfRQ.

256. Rosalind Hinde, "The Cnidaria and Ctenophora," In Anderson, D.T. (ed.). *Invertebrate Zoology*. Oxford University Press (1998): pp. 28–57. ISBN 978-0-19-551368-4.

257. Mark Q. Martindale and Jonathan Q. Henry. American Zoology., 38:672-684 (1998) The Development of Radial and Biradial Symmetry: The Evolution of Bilaterality

258. Paulette Cartwright, Susan L. Halgedahl, Jonathan R. Hendricks, Richard D. Jarrard, et al. (2007) "Exceptionally Preserved Jellyfishes from the Middle Cambrian," *PLoS* ONE 2(10): e1121. https://doi.org/10.1371/journal.pone.0001121

259. Eric Williams, July 4, 2006, Cape Cod Times. "Man-of-Wars close beaches," accessed August 14, 2022, https://www.capecodtimes.com/story/news/2006/07/04/man-wars-close-beaches/50891636007/.

260. Britannica, T. Editors of Encyclopaedia. "Portuguese man-of-war." Encyclopedia Britannica, May 3, 2020. https://www.britannica.com/animal/Portuguese-man-of-war.

261. André W. Sleeswyk, "Carvel-planking and Carvel Ships in the North of Europe," *Archaeonautica*. 14: (1998): 223–228 (224f.)

262. Catriona Munro, Zer Vue, Richard Robert Behringer, & Casey W Dunn, "Morphology and development of the Portuguese man of war, *Physalia physalis*," *Sci Rep* 9, 15522 (2019). https://doi.org/10.1038/s41598-019-51842-1

263. Mary N. Arai, A Functional Biology of Scyphozoa. London: Chapman and Hall. (1997) pp. 68–206. ISBN 978-0-412-45110-2.

264. Alexander Agassiz's Illustrated Catalogue of the Museum of Comparative Zoology North American Acalephae · Volume 2, pg 44

265. Eugene Kozloff, "Seashore Life of the Northern Pacific Coast," University of Washington Press. pp. 54, 56 (1983 edition). ISBN 978-0295960845.

266. Dale R. Calder, "Development of the sea nettle *Chrysaora quinquecirrha*," *Chesapeake Science*, 13 (1972): 40-44.

267. David G. Cargo, Glenn E. Rabenold. "Observations on the asexual reproductive activities of the sessile stages of the sea nettle *Chrysaora quinquecirrha* (Scyphozoa)," *Estuaries and Coasts*, 3 (1978): 20-27.

268. Paul Bologna, John Gaynor, Christie Barry, & Dena Restaino, "Top-Down Impacts of Sea Nettles (*Chrysaora quinquecirrha*) on Pelagic Community Structure in Barnegat Bay, New Jersey, U.S.A." Journal of Coastal Research. 78. (2017): 193-204. 10.2112/SI78-015.1.

269. Ame Jernelöv, "The Warty Comb Jelly in the Black Sea," In: The Long-Term Fate of Invasive Species. Springer, Cham. (2017). https://doi.org/10.1007/978-3-319-55396-2_18

270. Sandra Kube, Lutz Postel, Christopher Honnef, and Christina B. Augustin, "Mnemiopsis leidyi in the Baltic Sea - distribution and overwintering between autumn 2006 and spring 2007," *Aquatic Invasions*, 2 (2) (2007): 137–145. doi:10.3391/ai.2007.2.2.9

271. Ahmet Erkan Kideys, "Fall and Rise of the Black Sea Ecosystem." *Science* 297 (2002): 1482 - 1484.

The Kraken's Younger Brother (June 24-30, 2010; Author's archives) combined with Abundance of Squid (June 19, 2014)

https://yesterdaysisland.com/abundance-squid/

272.   Richard E. Young, Michael Vecchione, and Katharine M. Mangold, "Decapodiformes Leach, 1817. Squids, cuttlefishes and their relatives," In "The Tree of Life Web Project," (2008) accessed August 22,2022, http://tolweb.org/Decapodiformes/.

273.   Richard E. Young, Michael Vecchione, & Desmond T. Donovan, "The evolution of coleoid cephalopods and their present biodiversity and ecology". *South African Journal of Marine Science,* 20 (1) (1998): 393–420. doi:10.2989/025776198784126287

274.   Edward E. Ruppert, Richard S. Fox, Robert D. Barnes, Invertebrate Zoology (7th ed.). CEngage Learning. (2004): pp. 343–367. ISBN 978-81-315-0104-7.

275.   Michelle D. Staudinger, Ryan J. McAlarney, William A. McLellan, D. Ann Pabst, "Foraging ecology and niche overlap in pygmy (Kogia breviceps) and dwarf (Kogia sima) sperm whales from waters of the U.S. mid-Atlantic coast," *Mar Mam Sci,* 30 (2014): 626-655. https://doi.org/10.1111/mms.12064

276.   Gaius Plinius Secundus; Jean Harduin (commentator) (1827). "Ad Pliniam Vitam Excursus I: de Plinii Patria". Caii Plinii Secundi Historiae Naturalis Libri XXXVII. Bibliotheca Classica Latina (in Latin and French). Vol. 1. C. Alexandre; N.E. Lemaire (editors and contributors). Paris: Didot. pp. XLIX–L.Xxx

277.   Edward E. Ruppert, et al. Invertebrate Zoology. Pp 343-367.

278.   The Nobel Prize in Physiology or Medicine 1963. Accessed August 12, 2022.

http://nobelprize.org/nobel_prizes/medicine/laureate s/1963/

279.   Julian Finn, "Doryteuthis (Amerigo) pealeii (Lesueur, 1821)". World Register of Marine Species. Flanders Marine Institute. (2016)

280.   Brooke A. Lowman, Andrew W. Jones, Jeffrey P. Pessutti, Anna M. Mercer, et al. "Northern Shortfin Squid (Illex illecebrosus) Fishery Footprint on the Northeast US Continental Shelf," *Frontiers in Marine Science*. Volume 8 (2021): DOI=10.3389/fmars.2021.631657

281.   Nadav Shashar & Roger Hanlon, "Spawning behavior dynamics at communal egg beds in the squid *Doryteuthis (Loligo) pealeii*," *Journal of Experimental Marine Biology and Ecology* 447.(2013): 65-74. 10.1016/j.jembe.2013.02.011.

282.   Jon K.T. Brodziak and William K. Macy, III. Growth of long-finned squid, *Loligo pealei*, in the northwest Atlantic. Fish. Bull. (U.S.) 94 (1996): 212-236.

283.   Marine Biological Laboratory. "Newly discovered squid pheromone sparks extreme aggression on contact." *ScienceDaily*. www.sciencedaily.com/releases/2011/02/11021012292 6.htm (accessed August 13, 2022).

284.   Christian Ibáñez & Friedemann Keyl, "Cannibalism in cephalopods," *Reviews in Fish Biology and Fisheries*, 20. (2009): 123-136. 10.1007/s11160-009-9129-y.

285.   Helena Smith, "Coordinated Hunting in Red Devils," *Deep Sea News*, Accessed August 12,2022, http://deepseanews.com/2012/06/coordinated-hunting-in-red-devils/

286.   Richard Edward Young and Clyde Roper, "Bioluminescent countershading in midwater animals: evidence from living squid," Science. 191 (4231) (March 1976): 1046–1048. doi:10.1126/science.1251214

287. Ryan Gilmore, Robyn Crook, R, and Jacob L. Krans, "Cephalopod Camouflage: Cells and Organs of the Skin". *Nature Education.* 9 (2) (2016): 1.

288. Yung-Chieh Liu, Tsung-Han Liu, Chun-Chieh Yu, Chia-Hao Su, and Chuan-Chin Chiao, "Mismatch between the eye and the optic lobe in the giant squid," *Royal Society Open Science*, 4(7), (2017): 170289. https://doi.org/10.1098/rsos.170289

289. Alexander I. Arkhipkin and Vladimir V. Laptikhovsky, "Observation of penis elongation in *Onykia ingens*: implications for spermatophore transfer in deep-water squid," *Journal of Molluscan Studies*, Volume 76, Issue 3, August 2010, Pages 299–300, https://doi.org/10.1093/mollus/eyq019

290. National Oceanic and Atmospheric Administration (NOAA) Fisheries, "Longfin squid," accessed August 12, 2022, https://www.fisheries.noaa.gov/species/longfin-squid

291. National Oceanic and Atmospheric Administration (NOAA) Fisheries. "Longfin squid: Landings," accessed August 13th, 2022. Last updated August 11, 2022. https://www.fisheries.noaa.gov/foss/f?p=215:200:1716 8914882780:Mail:NO:::

292. Merriam-Webster, "Calamari." Merriam-Webster.com Dictionary, Accessed August 14, 2022, https://www.merriam-webster.com/dictionary/calamari.

About the Author

Sarah Oktay was born in a small town in Oklahoma. She is the Executive Director for the Center for Coastal Studies in Provincetown Massachusetts. From 2018-2021 she was the Director of Strategic Engagement for the Natural Reserves System at the University of California Davis. Prior to that, Sarah was Director of Institutional Advancement at the Rocky Mountain Biological Laboratory in Gothic, CO.

She received her B.S. in Marine Science and a Ph.D. in Chemical Oceanography from Texas A&M University - Galveston. From 2003-2016 she was the Executive Director of the University of Massachusetts-Boston Nantucket Field Station, a biological field station on Nantucket. Her research focused on climate change, carbon transport, and harbor water quality. After 9-11, she mapped the chemical signature of the World Trade Center ash and tracked it in the Hudson River. She is a board member and the 2020 President of the Society of Women Geographers, and she has been on the boards of many civic and nonprofit groups. She served as president of the Organization of Biological Field Stations, a organization representing several hundred field stations across the globe, from 2014-2016 and has been on their board for 13 years. Her nine years of service on the Nantucket Conservation Commission has been featured in *Vanity Fair, Yankee Magazine, Cape Cod Times,* ABC.com, CNN, the movie "Rising Tides", and many other news outlets. In 2020 she released a book of science-based poetry entitled "Sifting Light from the Darkness" via Noble Fir Press and is working on four books of nature essays. She contributed leadership and mentorship chapters to a March 2022 sociology textbook (Lexington Books) entitled "Women of the Wild: Challenging Gender Disparities in Field Stations and Marine Laboratories."

www.ingramcontent.com/pod-product-compliance
Lightning Source LLC
Chambersburg PA
CBHW072121270326
41931CB00010B/1627